Affordable Reliability Engineering

Life-Cycle Cost Analysis for Sustainability and Logistical Support

Affordable Reliability Engineering

Life-Cycle Cost Analysis for Sustainability and Logistical Support

William R. Wessels • Daniel S. Sillivant

CRC Press
Taylor & Francis Group
Boca Raton London New York

CRC Press is an imprint of the
Taylor & Francis Group, an **informa** business

CRC Press
Taylor & Francis Group
6000 Broken Sound Parkway NW, Suite 300
Boca Raton, FL 33487-2742

First issued in paperback 2017

© 2015 by Taylor & Francis Group, LLC
CRC Press is an imprint of Taylor & Francis Group, an Informa business

No claim to original U.S. Government works

ISBN-13: 978-1-4822-1964-7 (hbk)
ISBN-13: 978-1-138-74760-9 (pbk)

Visit the Taylor & Francis Web site at
http://www.taylorandfrancis.com

and the CRC Press Web site at
http://www.crcpress.com

To my O.A.O., and the love of my life, Tudor.

Bill Wessels

To my parents who were concerned with my writing ability when I

was younger. I would not have made it this far without them.

Daniel Sillivant

Contents

Preface

The book is written for engineers and managers. It presents procedures that use reliability information to determine life-cycle costs for part selection and logistical support analysis performed in system design and sustainment alternatives.

Organizations must be economically efficient to survive in the global competitive markets they serve. A reputation for high reliability and best costs of acquisition and life-cycle sustainability provides organizations with a competitive advantage that increases sales revenues and market share. Reliability-based life-cycle economic analysis enables an organization to be economically efficient and achieve a reputation for high reliability at best cost.

This book seeks to achieve two objectives:

1. Provide managers with an understanding of a reliability engineering program so that they can justify and fund reliability analyses
2. Provide engineers with an understanding of a reliability engineering program so that they can perform reliability analyses

Many engineers and managers believe that reliability is a discipline that requires specialists in the field. This book demonstrates that every organization already employs "reliability engineers" and "reliability managers." They are the organization's employees who have an intimate knowledge of the organization's products, capital equipment and machinery, culture, and customers. Engineers have a knowledge base from education and experience that is focused on achieving part functionality; reliability engineering uses the same knowledge base to focus on understanding part failure. Managers have a knowledge base from education and experience that is focused on allocation of scarce resources to implement part functionality; reliability management uses the same knowledge base to make cost optimum decisions to mitigate part failure.

Time-value of money is the essential principle for life-cycle economic analysis. Economic analysis requires information that estimates the magnitude of cost events and when the cost events will occur. This book shows how reliability analysis provides estimates of what the cost events will be, the duration of the cost events, and when the cost events will occur.

A final observation: All life-cycle sustainability costs for a system have a single cause—part failure. Absent part failure, an organization has no need for capital and operating investment to sustain a system. Understanding part failure in design and sustainment of fielded parts enables engineers and managers to optimize capital and operating investment.

About the Authors

Bill Wessels has over 40 years of experience in system design and sustainability. From 1975 to 1989, Bill worked as a field engineer for mining companies throughout the United States, Australia, Canada, Africa, and Europe to implement innovations in system sustainability for excavators, haul trucks, and process machinery. From 1989 to 2005, Bill worked as a reliability engineer for companies that performed research and design for aerospace payloads, defense aviation and weapons systems, chemical processes, and biomedical devices. Since 2005, Bill works at the University of Alabama in Huntsville, where he cofounded the Reliability and Failure Analysis Laboratory and performs basic and applied research in design-for-reliability, reliability-based maintainability, and reliability-based life-cycle economic analysis. Bill teaches professional development tutorials and consults in the United States and the Pacific Rim. He has a BS degree in engineering from the United States Military Academy at West Point, an MBA in decisions sciences from the University of Alabama in Tuscaloosa, and a PhD in systems engineering from the University of Alabama in Huntsville. He is a registered professional engineer in mechanical engineering and a certified reliability engineer. Bill and Tudor live on a small farm in North Alabama where they raise guard donkeys, chickens, worms, and dogs.

Daniel Sillivant is a researcher in the Research Institute at the University of Alabama in Huntsville (UAH) performing basic and applied research and investigations in reliability life-cycle modeling for aviation and sensors systems. He is published in peer-reviewed proceedings for the *International Mechanical Engineering Congress and Exposition*; *Reliability, Availability, Maintainability Workshop*; and *Industry, Engineering, and Management Systems*. Daniel has completed the requirements and has begun his dissertation research in reliability based life-cycle economic modeling for implementation of reliability-centered maintenance. He has a bachelor's degree in Chemical Engineering and a master's degree in Industrial/Reliability Engineering from UAH. His other certificates include Lean Concepts Training and Six Sigma Green Belt; and he passed the Fundamentals of Engineering Exam.

1

Scope of Reliability-Based Life-Cycle Economical Analysis

The objective of this book is to provide a framework to managers and engineers to develop and implement a reliability program for their organization that provides information that goes beyond verification that reliability requirements are met. Reliability analysis also provides information that defines life-cycle cost events that enables performance of engineering economic analyses. Engineering economic analyses investigate design and sustainability alternatives to enable engineers and managers to make life-cycle economic decisions under uncertainty. Engineering economic analysis is no different than financial analysis taught in business schools. Engineering economic analyses characterize

1. Cash flows over an evaluation period

   [Cash flows are cost estimates made for present, recurring, and future amounts.]
2. Equivalent financial metrics based on the time value of money

   [Equivalent financial metrics are present values of cash transactions in a specific time period, net present values for all present values, and equivalent recurring values of net present values.]

Engineering economic analysis applies to

1. Design of systems
2. Systems engineering and integration
3. Maintenance and sustainability of systems

It enables engineers and managers to determine the lowest life-cycle costs for part selection, design configuration options, implementation of maintenance practices, spare parts strategies, and logistical resources.

[Maintainability is defined as repair and logistics events associated with a system downing event. Sustainability is defined as maintainability and logistical support analysis.]

Implementation of reliability-based life-cycle economic analysis requires that

1. Managers understand methods and procedures that must be employed in a reliability program, and the cost benefit for investment in reliability analyses
2. Engineers understand the analytical investigations required within the context of a reliability program that provides value to management

This book is based on the premise that all system sustainment costs have a single cause: part failure. Not a plurality of system sustainment costs, not the majority of system sustainment costs—ALL! If parts do not fail then there will be no need for maintenance actions, no maintenance facility requirements, no spare parts requirements, and no logistical support requirements.

Background

An organization or consumer that owns and operates capital assets, a system, incurs three categories of costs: acquisition, operations, and maintenance. The operations and maintenance costs are referred to as O&M costs, also known as operations and sustainment. Operational costs include labor, materials, and overhead expenses for system functionality and servicing. Sustainment costs include labor, materials, and overhead expenses for maintenance and logistical support. Maintenance includes all events that are performed to

1. Restore a system to functionality following a system downing event
2. Prevent an unscheduled system downing event during scheduled system operation

Logistical support includes all events that provide the resources required to perform maintenance events. Maintenance personnel, tools, facilities, spare parts, specialty equipment, and contracted maintenance services are just a few items involved in logistical support events.

Servicing is not maintenance. It is the replenishment of expendable items: fuel, lubricants, and coolants, to name a few. Servicing is not a repair task and may be performed with equal skill by operators or maintainers. If an organization assigns system servicing to the maintenance organization it does not change the fact that servicing is an operational event. Similarly, although the costs of fuels, lubricants, and coolants are budgeted for production expenses, they are still operational expenses. Alice Roosevelt Longworth best described servicing as, "Fill what's empty, empty what's full, and scratch where it itches."

Reliability engineering is a discipline that investigates and analyzes part failure. It is a multidiscipline practice of engineering as it applies to all basic engineering disciplines: mechanical, civil, and electrical, and their sub-disciplines: chemical, nuclear, mining, aerospace systems, industrial, etc. As with all engineering disciplines, the application of reliability engineering falls in design and development, systems engineering and integration, and system sustainment.

Engineers who perform design and development are provided

1. System requirements, typically functional, that have been allocated down through a work breakdown structure to the lowest design hierarchical level, typically an assembly concept
2. A blank sheet of paper

They perform part design analysis that yields design art and a bill of materials, for the assembly concept. They create a tangible assembly that is comprised of parts from a concept. The objective of design engineers is to achieve functionality. Design for reliability performed by design engineers applies the same methods for design analysis that achieve functionality to then understand and model failure mechanisms and modes for part failure. It should be intuitively obvious that no one is better qualified or capable of analyzing the failure mechanisms acting on a part than the engineer who designed that part.

Engineers who perform system sustainment must understand

1. Part design analysis; although they do not perform, nor do they influence system design, they often evaluate third-party vendors for spare parts.
2. Systems engineering and integration; although they do not perform or influence system engineering and integration, they apply their understanding of design and systems engineering and integration to define and implement maintenance policies and practices that enable the O&M organization to achieve the optimum functional and economic performance of the system.

Sustainability engineers apply their understanding of part design and systems engineering to restore systems to full functionality following a downing event caused by a part failure. Reliability-trained sustainability engineers preserve system functionality, improve maintenance policy and practices, and optimize life-cycle economics of the system. The former is reactive; the latter is proactive. The former cannot influence system availability; the latter can.

Engineers who perform systems engineering and integration provide the design engineers with the allocation of the system requirements and the work breakdown structure, integrate the design analysis, design art and bill of materials, and determine whether the system requirement has been achieved. Systems engineers employ tests and evaluations that may return design modifications to the design engineers through a feedback loop, perform design reviews to assure system configuration of all lower hierarchy design into higher hierarchy design, and ultimately deliver the documentation that enables production of the system. Systems reliability engineering performed by systems engineers applies the same methods for allocation and integration of functional requirements to reliability requirements. The design and implementation of test and evaluation to verify that functional requirements have been met also provide data that can be used to fit reliability math models to characterize the reliability parameters of the design.

Reliability engineering is not, and should not be viewed as, a separate engineering discipline performed by specialty engineers, but rather it is a set of skills that should be applied by design engineers, systems engineers, and sustainability engineers.

The performance of reliability engineering has the same constraints that apply to design, systems, and sustainability engineering:

1. Technical constraints
2. Cost constraints
3. Schedule constraints

All engineering tasks performed by an organization are governed by the following principles:

- Cash flow is the lifeblood of an organization.

  No organization can operate in a negative cash flow scenario, whether the sources of cash are from revenue or debt. An organization with negative cash flow cannot pay its bills and will fail. Therefore the role of engineering tasks is to create value greater than the cost of performing the engineering tasks. The performance of reliability engineering tasks must add value to the organization. In design, that value can be manifested by more efficient design or by reliability requirements that improve the functional performance of the product, more efficiently utilize design budgets, and more efficiently achieve schedule requirements.
- "What one can measure one can control, and what one can control one can manage."

Bill Hewlett
Cofounder of Hewlett-Packard

The role of managers is to allocate scarce resources among infinite wants. Managers make their decisions under conditions of uncertainty. That uncertainty includes a lack of perfect information. The more information that a manager possesses, the better he or she can control and allocate the resources. Engineers are masters at developing and implementing metrics that can be used to enable managers to control design, systems engineering, and sustainability tasks.

- They would rather tolerate a problem they cannot solve than implement a solution they do not understand.

  Management practices recognize the need to effectively communicate the objectives of their organization to their subordinate staff. Unfortunately, too often engineers do a poor job of communicating the results of their analysis to the managers. The burden to effectively communicate the metrics an engineer develops and how those metrics can enable the manager to control their scarce resources is on the engineer, not on the manager. The performance of reliability tasks is a demand for resources made by the engineer on the manager. If the manager does not understand the value added from the reliability analysis, they are loath to approve the performance of reliability engineering tasks.

This book will serve to provide the engineer and the manager with an understanding of what reliability analysis is and how reliability analysis will add value to the organization that far exceeds its cost.

Mission Reliability, Maintainability, and Availability

Mission reliability is the probability that a part, an assembly, and a system will perform its function without failure for a stated mission duration under specified conditions of use. Mission reliability is characterized at the part level by fitting time-to-failure data to a failure math model. The part failure math model yields the part cumulative failure math model, the part survival function, the part mission reliability function, and the part hazard function.

Maintainability is the probability that a system will be restored to functionality following a part failure. Part maintainability is characterized by fitting time-to-repair data to a repair math model and fitting logistical downtime data to a logistical downtime math model.

Availability is the probability that a system will be able to begin a mission when scheduled and is the ratio of "uptime" to the sum of "uptime" and "downtime." Availability is a function of reliability, "uptime," and maintainability, "downtime." Availability takes three forms:

1. Inherent—a design predictive metric
2. Operational—a design and sustainment predictive metric
3. Achieved—a sustainment management metric

Reliability engineering is practiced by organizations that design, develop, test, evaluate, and manufacture systems. Or rather, organizations implement reliability engineering as they think it should be performed with little agreement on what reliability engineering is. A common approach is failure modes and effects analysis equals reliability engineering. Mean time between failure, or its corresponding failure rate, equals reliability engineering is another common approach. Expressing a system's reliability as either a percentage or a mean time between failure is yet another approach. Describing reliability engineering is similar to the anecdote about ten blind men describing an elephant based on the body part they grasped in their hands—the tail, leg, ear, and trunk. Many organizations eschew reliability engineering because it is too expensive and takes too much time; being first to market and fixing system problems in the field has worked to their benefit for decades. A few organizations truly understand reliability engineering. Caterpillar Tractor is one such company, and they have built a well-earned global dominant market share for mining and construction machinery due in part for their demonstrated system reliability.

Reliability History

Reliability is a young discipline. It emerged between World War II and the Korean War as military aircraft changed from propeller to jet propulsion. Reliability engineering was developed as a way to understand the application of digital and electronic technology that enabled the leap from propellers to jets. The reliability field was dominated by physicists, electrical engineers, and statisticians. In 1952, the Institute of Electrical and Electronic Engineers formed the first reliability engineering technical society, and a discipline was born. Academe was slow to respond. The University of Maryland and the University of Arizona were the first to offer a graduate major/minor option in reliability in the 1980s, and there are fewer than twenty-five university programs worldwide by 2014. The dominance of the electrical engineering discipline in the education and practice of reliability engineering is manifest by the absence of a reliability division in the American Society of Mechanical Engineers. A Google search for reliability engineering books claims only 850 listings, compared to tens of thousands of mechanical, electrical, and civil engineering books. The Reliability and Maintainability Symposium, the oldest and most respected international annual gathering of reliability engineers, attracts 600 attendees representing reliability divisions from eight technical societies, compared to the 3,500 attendees for the annual International Mechanical Engineering Congress and Exhibition.

The timeline in Figure 1.1 shows the relative maturity of the engineering disciplines. As can be seen in the top timeline, civil engineering has been in practice for over 5,000 years. It is unquestionably the most mature

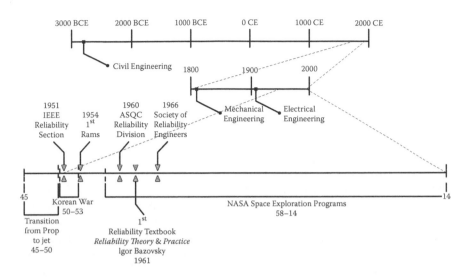

FIGURE 1.1
Reliability engineering timeline.

engineering discipline, is well understood by all competent practitioners, and experiences very slow innovation. The middle timeline shows the emergence of mechanical engineering from the industrial revolution approximately 200 years ago. It too is a mature engineering discipline; past practices are well understood by all competent practitioners; however, it is experiencing technological innovation that requires continual professional development on the part of the engineer. Electrical engineering became a recognized discipline in the last 100 years. It can be said to be a maturing engineering discipline; past practices for power transmission in alternating and direct current are well understood by practitioners; however, innovation in electronic and digital equipment is advancing very rapidly and requires extensive constant professional development in order to keep pace with that innovation.

The bottom timeline shows that reliability engineering is a very recent phenomenon. There is ample evidence that it has yet to reach maturity. As mentioned previously, the concepts are still evolving. The body of knowledge is nascent, and its application to design and sustainability of systems is not well accepted. Managers seldom recognize how reliability investigations could affect the value of their product; they find the test procedures to be expensive with limited benefit; and they are not quite certain how to use the information gathered from reliability investigations.

This book seeks to resolve that uncertainty and to establish the value of reliability engineering.

Reliability Engineering Approaches

The reliability engineering discipline is comprised of three analytical approaches:

Mission Reliability

Mission reliability is the probability that a part will function without failure for a specific mission duration under stated conditions of use. It is a conditional probability that the part will survive the next mission given that it has survived to a specific time in its useful life. It is often stated as a percentage, for example, a part is stated to have a mission reliability of 99%. Often the mission reliability is expressed in mean time between failure, for example, a part is stated to have a mean time between failure of 300 hours. A reliability investigation for mission reliability will use analysis and test data to fit a failure math model, a cumulative failure math model, a survival function, a mission reliability function, and a hazard function. Part mission reliability math models are fit by design engineers using design analysis, test, and evaluation, and by sustainability engineers using field failure data from operations.

Maintainability

Maintainability is the probability that a part will be restored following its failure. Maintainability is typically expressed as a mean time to repair, a mean maintenance time, or mean downtime. A reliability investigation for maintainability will use analysis and test data to fit a repair math model and a cumulative repair math model and will use logistical downtime models that characterize the common cause of delay between a failure of a part and a maintenance action followed by the completion of the maintenance action and the system's return to service. The two logistical downtime models are referred to as prerepair and postrepair logistics downtime. The mean downtime math model is the sum of the mean time to repair, or the mean maintenance time, and the two logistical downtime math models. Part maintainability math models can be estimated by design engineers in laboratory repair experiments but are more relevant when performed by sustainability engineers under their specific conditions of use.

Availability

Availability is the probability that a part will be capable of performing its mission in time. Availability is expressed as the inherent availability, a design criterion; operational availability, a predictive metric; and achieved availability, a management metric. Availability is the ratio of uptime divided by uptime plus downtime. The variations between inherent, operational, and achieved availability are determined by the definition for uptime and downtime.

The inherent availability is calculated as the mean time between failure divided by the sum of the mean time between failure and the mean time to repair. It can be estimated by the design engineer, although the mean time to repair will not necessarily reflect the actual experienced mean time to repair by the user. Its value, however, can be viewed as the optimum availability that the part can achieve.

The operational availability is calculated as the mean time between failure divided by the sum of the mean time between failure, the mean time to repair, the prerepair logistics downtime, and the postrepair logistics downtime. Design engineers can characterize the operational availability with estimates for logistical downtime; however, the sustainability engineer can calculate the operational availability based on the best information available on what the actual logistics downtime will be.

The achieved availability is calculated as the mean time between failure divided by the mean time between failure plus the mean downtime. This calculation is performed for a prior reporting period and is specific only to the duration of that period. For example, it may be calculated for the previous month, the previous quarter, or the previous year. It can only be characterized by sustainability engineers.

Availability math models are difficult for design engineers to fit because they lack the understanding of the logistical downtime conditions for the system user; however, they can be very accurately calculated by the sustainability engineer for their systems in use.

Mission reliability, maintainability, and logistics downtime math models describe the operation and maintenance cycle for a part as illustrated in Figure 1.2. Failure math models describe the behavior of part failure and the likelihood of the time region in which a part failure will occur. Prerepair logistical downtime models describe the expected delay time between the occurrence of the part failure and the commencement of the maintenance action. Repair math models describe the expected time to repair for the part. Postrepair logistics downtime models describe the expected delay time between the completion of maintenance action and the return of the system to service. The mean downtime is the sum of the prerepair logistics downtime, the time to repair, and the postrepair logistics downtime. The operation and maintenance cycle for parts determine the availability of the assembly and all higher design configurations through the system level. This information provides metrics to the manager that determine what parts will fail, when the failure will occur, and how long the system will be in a down state.

Two barriers to effective implementation of reliability engineering and organizations have been previously mentioned, ineffective communication of what reliability metrics mean to the organization and lack of understanding by the manager of the value added by reliability engineering. Both

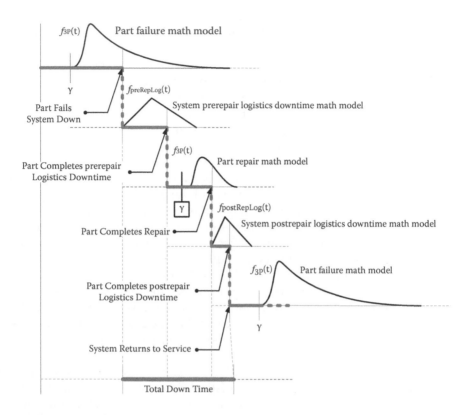

FIGURE 1.2
Downtime timeline.

barriers are reinforced by commonly observed organizational structure. The solution to eliminating these barriers can be found in revising the organizational structure.

Organizational Structure

Figure 1.3 illustrates the typical organization structure for organizations that claim to implement reliability analysis. The responsibility for design, systems engineering, and sustainment engineering to meet the functional requirements for the system are separate from the responsibility to implement quality, reliability, and safety engineering. Reliability engineering is relegated to a specialty engineering cubbyhole that is isolated from the performance of design, systems, and sustainability engineering. That isolation can range from separate offices in the same building, to separation in different buildings, to separation in different cities or states. Often the role of the reliability engineer is to audit work performed by design, systems, and sustainment engineers after the work has been performed—when it is too late to implement their findings. The situation is often exacerbated by a tension between

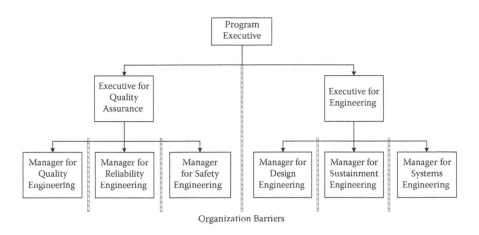

FIGURE 1.3
Organizational structure.

the organizations. Often there is only a single reliability engineer assigned to a project in which multidiscipline design, systems, and sustainment engineering functions are performed. Such an engineer is not really qualified to address an engineering discipline in which they are untrained or not experienced. Compound the detachment between the personnel with the lack of understanding of what constitutes a reliability program and it is no wonder that managers do not see the value added to the reliability investment.

Too often there is a lack of understanding between the distinctions for reliability and quality. They are often viewed as being the same thing because they share common analytical tools. For example, failure modes, effects, and criticality analyses (FMECA) are performed by quality, reliability, and safety engineers. In fact they couldn't be more different. The role of quality is to assess whether a process is in control and capable, to assess the effectiveness of part selection, to audit vendors for quality of the parts they deliver, and to evaluate the output of the process that manufactures the part. Reliability addresses the failure mechanisms that act on the part, the failure modes, damage caused by the failure mechanisms, and the effects of the failure modes on the part, on the assembly in which it is integrated, and on the system. Safety addresses the hazards to people and systems posed by the design and operation of the part. Therefore, while each uses a common tool, FMECA, each applies the tools to different objectives: quality evaluates the manufacturing process, reliability evaluates the part design, and safety evaluates the consequences of system operation.

Many organizations try to implement management initiatives that seek to eliminate the barriers between specialty engineering and their functional engineering departments. An example, one of many, is concurrent engineering. Figure 1.4 illustrates a notional teaming structure for a concurrent engineering approach. The basic idea in concurrent engineering is to involve

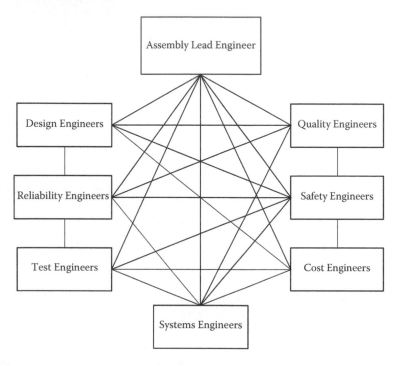

FIGURE 1.4
Revised organizational structure.

the specialty engineers on the design, systems engineering, or sustainment engineering team. Unfortunately this does not alleviate the problems previously mentioned. The specialty engineering role is still one of auditing the functional engineer. There is still just one reliability engineer with a single engineering discipline dealing with multidiscipline design, systems, and sustainability engineers. It places the reliability engineer in the role of looking over the functional engineer's shoulder rather than participating directly in the design.

This book proposes a revised organizational structure illustrated in Figure 1.5. In the revised organization the functional engineers are trained to perform the appropriate reliability tasks for design, systems, and sustainability engineering. The functional engineers are the most qualified individuals to perform reliability tasks. The feedback loop is eliminated, since they will perform the reliability analysis as they are performing the functional analysis. The specialty engineering organization does not go away. It changes its mission from performing reliability analysis and auditing functional tasks to providing the training that is required for the functional engineers to perform the reliability investigations they require to meet the needs of the organization. The metrics developed by the functional engineers can be articulated to the engineering manager and the higher levels of

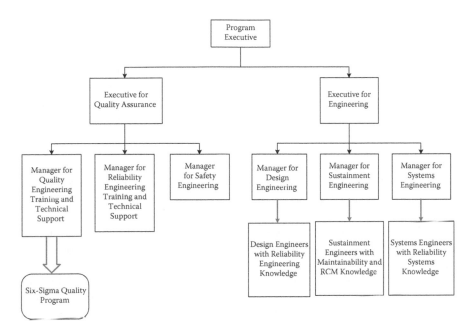

FIGURE 1.5
Organizational structure with reliability engineers.

management more effectively. The communication from management down to engineering is directed to the functional engineers. There is already a successful implementation of this organizational change—Six Sigma quality. Functional engineers use Six Sigma to perform quality analyses that enable them to achieve the quality goals for the production of the product. The specialty engineering organization provides training to the functional engineers to assure that they can perform these tasks. There is no reason that the same approach cannot be applied to the reliability engineering discipline in an organization.

Reliability Engineering Economics

Economic decisions are based on the perception of system affordability by the organization that operates and maintains a system. A system is affordable if and only if the acquisition, operation, and sustainment costs are less than the revenue from operating the system. A system is affordable to the consumer if and only if the price of the system is less than the perceived utility of the system.

Research and Development

Research and development is initiated by the perceived need for a system. Research organizations invest money in exploring how such systems can be developed and perform the proof of principle. That exploration includes not only the functionality of the system but also the economic feasibility of the system. Basic research investigates innovations that add to the body of knowledge for the engineering discipline; they create something that has never been done before. Examples of basic research and development are pharmaceutical companies that investigate and develop diagnostic and treatment devices, laboratories that formulate innovative materials, and organizations that conceptually formulate innovative products—the better mouse trap.

Applied research relies on existing technology and processes to develop an innovation in a product rather than adding to the body of knowledge. Examples include application of known best practices in automotive and digital technology to develop a smart car, best practices in nuclear technology and diagnostic equipment to develop a cancer treatment device that is nonintrusive, and applications for innovative materials to develop stronger lighter structures for aviation systems.

All research and development is constrained by technological and economic feasibility. Economic feasibility dictates that the ultimate commercial application of the research will generate a positive cash flow for the consumer of the technology. That is, can the customer afford to acquire and sustain the innovation? Part failure determines the sustainability affordability.

Design and Manufacture

Design and manufacture is the source of system quality and reliability. Design is the origin of all sustainability costs—part failure. Design is guided by requirements specifications that emphasize system functionality and constraints. Commercial reliability issues are typically concerned with warranty and safety issues, maintenance procedures and manuals, and avoidance of single points of failure for critical design configurations. The reliability issues of government agencies (defense, National Aeronautics and Space Administration, energy, transportation) often include system reliability requirements stated as mission reliability or mean time between downing events, mean time to repair, and availability.

Transportation and Distribution—The Supply Chain

Systems travel through a complex path, the supply chain from the manufacturer to the consumer. This path involves the transportation, handling, and storage of the system from the manufacturer to various distribution points that culminate with the retail transaction with the consumer. For example, consider a hydraulic cylinder actuator. The technology for a specific application is developed through applied research of alternatives. Results of the applied research yield engineering design. Engineering design is subject to test and evaluation. Final engineering design results in a manufacturing configuration. The part is then produced by a manufacturing process. The manufacturer stores its inventory prior to shipment. The product is packaged and shipped through a distribution chain that may include several intermediate organizations that handle, store, and reship the product. The product is acquired by the end user. The end user puts the product in its inventory, and finally the product is installed on the system for which it was designed.

Transportation and distribution provide an interesting demand on reliability analysis. On one hand, transportation and distribution demand reliable systems to perform their tasks. On the other hand, transportation and distribution provide sources of failure mechanisms (Figure 1.6) that act on the system that may not have been anticipated by the designer. The applied research and engineering design is focused on the actual operation of the product in its end state. Design and reliability analysis focuses on the operational and ambient conditions of use that provide the sources of failure mechanisms that will act on the product. Yet the transportation, material handling, and storage of the product provide failure mechanisms that will act on the product before it is installed on the end system.

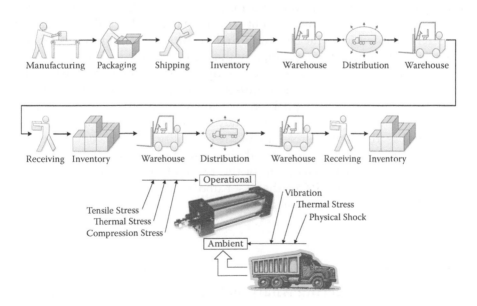

FIGURE 1.6
Asset failure mechanism sources.

Operation and Maintenance, O&M

The economic analysis for design and field sustainment is based on cash flow. Cash flow has three sources:

1. Owners' equity
2. Debt
3. Revenue

Cash flow has three uses:

1. Acquisition of capital assets
2. Service on owners' equity and debt
3. Payment for direct and indirect expenses

Consumers can be either organizations or individual consumers. Organizations that acquire systems to perform work or provide a service are viewed as end users.

Economic models employ the economists' concept of the rational consumer. A rational consumer evaluates the acquisition cost as well as the operating and sustainment costs.

All organizations, whether they perform research and development, design and manufacture, or transportation and distribution, are consumers. Examples of end users would be a mining company that acquires a haul truck, a transportation company that acquires a material handling process, a research and development company that acquires a milling machine, and a design and manufacturing company that acquires process machinery.

A return on investment is not necessarily the evaluation criteria when the end user is a person, one who acquires a car, appliance, or toy. Such end users typically base their acquisition decision on the acquisition cost, and the utility that the asset provides them. They will make subsequent acquisition decisions based on their perception of the reliability of the system. They will employ word-of-mouth to express their satisfaction or dissatisfaction with their colleagues. Their perception of the reliability will be both tangible—in terms of the useful life of the asset and its cost of maintenance—and intangible—in terms of general dissatisfaction.

Reliability Engineering Analysis Impacts on Life-Cycle Costs

Organizations tend to be myopic when they evaluate the need and the value added of reliability engineering analysis. Too often the approach taken by management is to apply reliability analysis solely for the purpose of determining whether a part meets or exceeds the reliability requirement for the system. Part reliability analysis is expensive relative to the design analysis, the systems engineering and integration, and the sustainability costs. And too often the reliability requirement is viewed as an abstraction that managers do not understand. Reliability-based economic analysis can evaluate the life-cycle costs for part selection in design. And it can be applied to evaluate life-cycle costs for implementing changes to maintainability practices (i.e., preventative maintenance, reliability-centered maintenance, etc.), optimizing sparing strategies and demand for specialty tools in field sustainment. Reliability-based analysis can also be applied to evaluate capital investment decisions for asset acquisition and facility design.

Engineering economic analysis addresses the time value of money. Reliability failure analysis provides an understanding of the occurrence of part failure—the demand for a maintenance action and when it will occur. Reliability failure analysis also provides an understanding of the demand function for spare parts and the time in service at which the hazard function, the instantaneous failure rate, equals the risk threshold that an organization is willing to accept for the likelihood of part failure on the next mission. The risk threshold enables an organization to evaluate the consumed life of a part until such time that it is economically feasible to remove and replace that part to prevent an unscheduled downing event of the system. Reliability maintainability analysis provides an understanding of the duration of logistical and repair time—the demand for logistical and maintenance assets, skills, materials, and overhead.

Reliability failure and maintainability analysis provides an understanding of all of the logistical events that result in the total downtime that the system experiences following a demand for maintenance so that it can be well understood and improved upon to optimize the downtime.

Typically organizations evaluate the cost of sustainment as the measurable expenses for labor, materials, and overhead. There is another cost associated with system downtime—lost opportunity cost. The lost opportunity cost is the revenue the organization forfeits because the system is in a down state. The lost opportunity cost often exceeds the direct expenses of a maintenance event by one or more orders of magnitude.

System downing events fall into two categories: unscheduled and scheduled. Unscheduled maintenance actions require the organization to react under conditions of uncertainty. Often the total downtime is not determined by the requirement to restore the system, but rather to await access to recovery vehicles, access to the maintenance facility, access to the required maintenance skills, and availability of spare parts. Scheduled maintenance actions eliminate the special cause variability from lack of resources because the resources can be planned for and marshaled at the appropriate time that the maintenance action is scheduled. Scheduled maintenance actions will reduce the common cause variability of logistical and maintenance actions. Both reduce downtime.

Consider the example of a mining haul truck. A critical part for the functionality of a mining haul truck is the air brake canister for the pneumatic braking. Any unscheduled maintenance action for the air brake canister will result in a system downing event that may or may not have catastrophic consequences. Catastrophic consequences might occur if the haul truck is backing up to the mine to dump overburden into the mined-out section. Brake failure causes the haul truck to back over the mine wall, destroying the haul truck and killing the operator. More likely the air brake canister failure will result in the haul truck being in a down state where the failure occurs. Murphy's Law would dictate that this would be with a loaded haul truck in the pit that cannot be recovered and brought back to the maintenance facility. In such a case the maintenance actions would involve doing the fault detection, fault isolation, and the repairs at the haul truck location regardless of the weather conditions or the time of day. Consider the scenario where the haul truck's unscheduled maintenance action occurs at 2 o'clock in the morning on a February winter day where the temperature is below freezing and sleet and snow are falling. The time that it takes to perform the repair is going to be longer than if the haul truck was in the maintenance bay under clean and environmentally controlled conditions. The prerepair and postrepair logistical downtime will be at a maximum. A reliability analysis of the air brake canister provides a sustainability engineer with the understanding that the consumer life of the part will exceed the allowable risk to commence the next mission after 1000 engine hours. As the haul truck approaches 1000 engine hours, maintenance schedules a maintenance action with the operations management in order to pull the truck from service. The truck is then brought into the maintenance facility, where the resources have been gathered to perform the repair event under ideal conditions. The time to repair will be significantly less. Prerepair and postrepair logistics downtime will be at the minimum duration for events that are required because the event has been scheduled.

Consider the organization that designed the haul truck. The engineering team that was responsible for the air brake canister was confronted with a variety of design alternatives for the selection of the part. Vendor information or reliability analysis performed on the alternatives determines that there are three parts that meet the reliability requirement as well as the functional requirements for the air brake canister. At this point the reliability analysis is terminated, and the discriminator will be the acquisition cost of the air brake canister. The reliability analysis that was performed to determine that the three alternatives met the reliability requirement can be combined with life-cycle economic analysis to evaluate the costs over the useful life of the part. The acquisition costs, sustainability costs, and the occurrence of those costs can be evaluated to determine which of the three parts has the lowest life-cycle cost to the end user. The reliability-based life-cycle economic analysis will determine the part that has the lowest life-cycle cost.

Reliability analysis goes beyond the evaluation of failure, repair, and logistical math models. Systems are comprised of thousands of parts, but a criticality analysis of the design enables the engineering organization to rank order the parts by criticality scale. The criticality scale enables management to determine which parts will be worthy of the investment in reliability investigation and analysis. The thousands of parts are now reduced to a few hundred. Rather than a single reliability engineer trying to build a system reliability model by developing and integrating the reliability models for the hundreds of parts, the revised organization structure will assign those hundreds of tasks to the engineers responsible for the design of the parts. Simulation software that performs reliability math modeling for systems is available to then integrate all of the parts into a systems model that enables higher design hierarchies to be evaluated for reliability maintainability and availability as well as the life-cycle analysis for the part selections that have been made by the design engineer.

This book provides the methodology for integrating part reliability, failure, maintainability, and logistic math models with their respective economic analysis into assembly reliability models. The assembly reliability models are then integrated into the work breakdown structure up to the system reliability model.

Implementation of Reliability Affordability

Implementation of reliability affordability is a four step process: (1) development of the critical items list; (2) performance of the qualitative failure, repair, and logistics analysis; (3) performance of the quantitative failure, repair, and logistics analysis; and (4) performance of the reliability-based life-cycle economic analysis, as illustrated in Figure 1.7.

The very first step in any reliability analysis is to develop a critical items list for the system. The objective for the critical items list is to identify the relevant few part failures that matter to the organization. It is a metric developed by engineers and managers that enables an organization to evaluate where to invest their scarce resources.

Criticality Analysis

Evaluation of criticality is more intuitive than analytical. Many analytical approaches for scoring a criticality metric exist; some have been promulgated

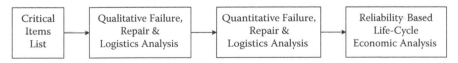

FIGURE 1.7
Reliability affordability process.

as standards by organizations and those that exist solely to write standards. Yet each analytical approach has a "subjective" element. One of the most well-known criticality metrics is the risk priority number (RPN) method. RPN is the product of

1. Risk of part failure (on a scale of 1, least risk, to 10, highest risk [but what risk?])
2. Probability, or likelihood, of part failure occurring (on a scale of 1, least likely [or zero chance?], to 10, certain likelihood [or near certain—depends on the wording of the standard])
3. Detectability of occurrence of part failure (on a scale of 1, least detectable—or not detectable, to 10, absolutely detectable)

RPN scores can range from 1 (1 × 1 × 1) to 1000 (10 × 10 × 10). The scoring approach is too subjective. Two or more engineers can arrive at scores that differ by an order of magnitude given the same part and the same understanding of the part.

Modal Criticality Number, C_m

The modal criticality number, C_m, is a criticality analysis method that is touted as a quantified metric. It assumes that a part has two or more failure modes. The modal criticality number is the product of

1. Failure model distribution, α

   Proportion of times a failure mode occurs. Assume a part has three failure modes, the proportion of occurrence of each failure mode is calculated from analysis of empirical information. The sum of the proportions must be 1.00.

2. Conditional probability that the failure mode results in part failure, β

   Likelihood that each failure mode actually causes part failure, ranging from 0.1, least likely, to 0.9, most likely, including increments of 0.1 in between, calculated from analysis of empirical information.

3. Part failure rate, λ

   Inverse of mean time between failure calculated from analysis of empirical information, typically expressed as failures, r, per million hours, $r \times 10^{-6}$.

4. System mission duration, τ, expressed in hours

Severity Analysis

Severity, or hazard, analysis is based on a safety engineering analysis that has gained acceptance by reliability engineers. Two metrics are scored on an X-Y grid

1. Likelihood of occurrence

   Use of subjective scale like the RPN measure of probability of occurrence, part failure rate, etc., depending on the standard.

2. Consequences of part failure

   Descriptors of the consequences, as few as four (catastrophic, major, minor, none) to ten or more, depending on the standard.

Severity analysis adds color coding for unacceptable criticality (red), acceptable criticality (green), and needs investigation (yellow).

All of these methods have practical limits to an organization.

- Each must follow a failure modes and effects analysis (FMEA). That means that a FMEA must be performed on 100% of the bills of materials. The amount of work required is time consuming and expensive.

- The modal criticality number and severity analysis requires a failure rate that must be acquired from empirical information, historical data, or experimental data. The investment in part failure analysis is performed absent the need. After all, criticality analysis is supposed to limit the number of parts that need investment in reliability investigation.

- Subjectivity and complexity of the process renders the findings suspect.

- Criticality can include a broader definition than part failure effects on the system. A part failure can have catastrophic consequences to an organization without any effect on the system's functionality— failure of parts required to meet safety and environmental regulatory compliance can lead to fines, shut downs, and criminal proceedings.

- High cost, long duration maintenance events for parts with very high mean time between failure demand every bit as much reliability investigation as parts that have catastrophic or operational failure consequences.

Reliability engineers must keep their eye on the goal—criticality analysis reduces the total number of parts in a system design configuration to the meaningful few, where meaningful is defined as economic impact to the organization.

The recommended procedure for developing the design critical items list (CIL) is to

1. Define the system design configuration through all hierarchical levels to the candidate critical parts

   Design criticality analysis demands understanding the context of the part in the system design hierarchy and within the assembly design configuration. The functional fault tree approach is an accepted method to trace system downing events through the design hierarchy to the candidate critical part.

2. Perform a criticality analysis for the candidate parts

   The design criticality analysis procedure presented in this book is based on John Moubray's consequences analysis from his book *Reliability-Centered Maintenance*. It has been modified to reflect system design and sustainment. The criticality analysis is performed from the vantage point of both the operator and the maintainer. The consequences analysis evaluates the context of part failure on the system as it relates to safety, operational performance, and regulatory conformance. The perception of failure addresses whether the part failure mode is evident or hidden to the operator or the maintainer. The P-F interval is the amount of time between the perception of the failure mode and the occurrence of the part failure state—the reaction time for the operator and the maintainer. The findings of the consequences analysis and perception of failure enables the organization to score part failures and rank them in the critical items list. The logic flow diagram for this procedure is presented in Figure 1.8.

3. Document the findings of the criticality analysis in a criticality analysis database

   The findings of the criticality analysis are entered in a criticality analysis database.

The procedure for developing the sustainability critical items list is more straightforward. Organizations that operate and maintain systems have a history of part failures that affect productivity and safety of their operations. Therefore they begin with a list of candidate critical parts for which they perform a failure report, root cause failure analysis, and corrective action analysis.

Qualitative Failure, Repair, and Logistics Analysis

Qualitative failure repair and logistics analysis is performed for critical parts only.

The objective of the qualitative failure, repair, and logistics analysis is to define the metrics that must be acquired through historical data, analysis, or experiment. All data are expensive and time consuming. Whether they are acquired through historical records, analysis, or experiment, a thorough understanding of what will be measured and how critical information is for managers so they can determine the impact of a reliability investigation on their budget and schedule. The qualitative failure, repair, and logistics analysis is the essential first step in the design of any reliability experiment.

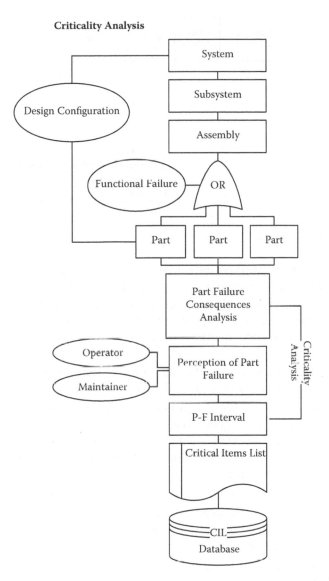

FIGURE 1.8
Critical items list logic.

The procedure for the qualitative failure repair logistics analysis is performed in 11 steps, as shown in Figure 1.9.

1. Identify the sources of failure mechanisms that act on the critical part.

   The sources of failure mechanisms include transportation, storage, and system functionality. System functionality includes operational and ambient conditions of use.

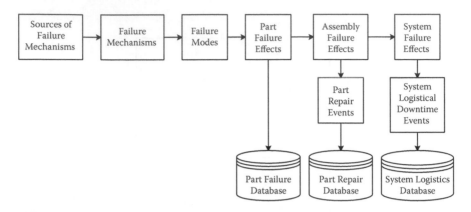

FIGURE 1.9
Qualitative reliability analyses.

2. Define the failure mechanisms that act on the critical part with metrics.

   Failure mechanisms are stresses or reactive agents, including but not limited to mechanical loads, thermal loads, and exposure to humidity, chemicals, and biological entities. Metrics are the measure of the magnitude and direction of the stress or potency of the reactive agent, for example.

3. Define the failure modes, damage done to the part, for each failure mechanism acting on the critical part with metrics.

   Failure modes can be categorized as change in shape of a part, *elongation*; change in geometry of a part, *crack initiation*; or change in part material properties, *corrosion*. Metrics are the measure of the strain, number, or length of the cracks, or the visual observation of corrosion, for example.

4. Define the part failure effects resulting from each failure mode.

   Part failure effects describe the symptoms of the damage and are expressed as misalignment of gear teeth, inability to withstand forces, for example.

5. Document the failure mechanism, modes, and effects in a part failure database.

   The findings of the failure mechanism, failure modes, and failure effects will be recorded in either a computational or relational database.

6. Define the assembly failure effects resulting from each part failure effect.

   Assembly effects describe the symptoms of the part failure effect on the assembly and tend to describe the specific loss of functionality, capability, or utility of the assembly.

7. Define the part repair events that must be performed on the assembly resulting from the part failure.

   Part repair events are the specific tasks that must be performed to remove and replace the failed part and are often characterized as work instructions. Part repair events also address the skill sets of the maintenance personnel, maintenance resources, materials, and overhead resources that are required to perform the repair event.

8. Document the part repair events in the part repair database.

   The findings of the part repair analysis will be recorded in either a computational or relational database.

9. Define the system effects resulting from each part failure effect with metrics.

   System effects describe the symptoms of the part failure effect on the system. Often times it tends to describe the operational and safety impact of the system.

10. Define the system prerepair and postrepair logistics events that must be performed prior to and following the repair maintenance events.

    Prerepair logistics events are the specific tasks that are performed from notification of the part failure to the commencement of part maintenance. Prerepair logistics events include transportation of the system to the maintenance facility, or transportation of the maintenance capability to the system, waiting time for facilities, specialty tools and labor, and resources required to perform these tasks.

    Postrepair logistics events are the specific tasks that are performed from the completion of part maintenance to the return of the system to service. Postrepair logistics events include system check out, documentation, recalibration, and transportation to return of service.

11. Document the system logistical downtime events in the system logistics database.

    The findings of the logistics analysis will be recorded in either a computational or relational database.

Quantitative Failure, Repair, and Logistics Analysis

Quantitative failure, repair, and logistics analysis is performed on the critical parts based on the findings of the qualitative failure, repair, and logistics analysis.

The quantitative failure analysis for part design begins with the system reliability requirements allocation for the reliability of that part. Part reliability allocations are performed by systems engineering. Part reliability allocations are goals that enable the design engineer to determine whether the design meets the system requirements.

Quantitative failure analysis for part sustainability by the user is guided by the realized reliability of the part over the fielded use of the system. The realized reliability serves as the goal for sustainability engineers to improve the system reliability.

Quantitative failure analysis uses the information in the failure database from the qualitative failure analysis to design and perform reliability experiments to fit part failure math models. Two reliability failure math models are presented in this book: time to failure and stress based. Time-to-failure math models are the conventional approach to reliability failure modeling. Stress-based failure math models recognize that time does not cause failure in mechanical design, stresses cause part failure.

It bears noting that the most important outcome from failure math modeling is the part hazard function (Figure 1.10). The part hazard function provides the information needed to understand the true behavior of part failure and to implement a preventive maintenance program. Specifically, the hazard function characterizes the risk-based consumed life of the part, where the risk, r, is the allowable risk of part failure that an organization is willing to accept for initiation of the next mission.

The quantitative part repair math model uses the information in the part repair database from the qualitative repair analysis to design and perform maintainability experiments to fit the part repair math models.

The quantitative logistics downtime math model uses the information from the part qualitative logistics downtime analysis to design and perform subjective Delphi surveys to fit the logistical downtime math models.

The findings of quantitative failure, repair, and logistics downtime analyses are documented in the respective databases. The logic flow diagram for the quantitative failure, repair, and logistics downtime analyses is provided in Figure 1.11.

The logic flow diagram describes parallel paths that are performed concurrently and by different engineering disciplines.

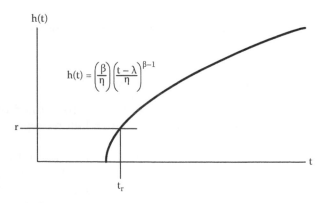

FIGURE 1.10
Hazard function.

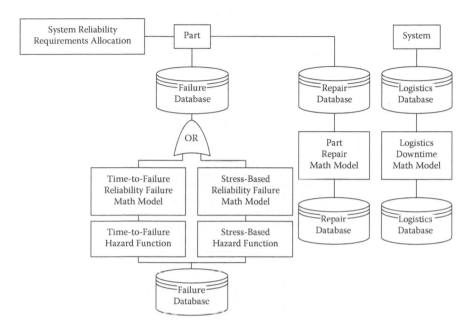

FIGURE 1.11
Quantitative reliability analyses.

Part quantitative failure math models provide only a point estimate and lower confidence limit for the mean time between failure for the part. They cannot address the behavior of the part over the useful life of the system. The same observation is true for the part quantitative repair and logistics downtime models. Yet, system life cycles experience part failures within all assemblies over the system's useful life, resulting in repetitions of the operations-maintenance cycles that are unique to each part. Assembly downing events caused by part failure impact the system's availability and life-cycle economics.

Reliability-based life-cycle economic analysis is performed using reliability simulation.

Life-cycle reliability simulation is performed for the part to evaluate part selection and the demand functions for logistics and repair events. Part failure, repair, and logistics math models are inputs to the life-cycle reliability simulation. Part selection is performed by design engineers and managers for make-buy decisions and by sustainability engineers and managers for O&M or third party vendor buy decisions. Life-cycle reliability simulation is performed for assemblies to evaluate the impact of part selection decisions on maintainability and availability and to influence the assembly design configuration. Life-cycle reliability simulation is performed for systems to evaluate the impact of part selection decisions and assembly design configuration on system availability. The logic flow chart for life-cycle reliability simulation is provided in Figure 1.12.

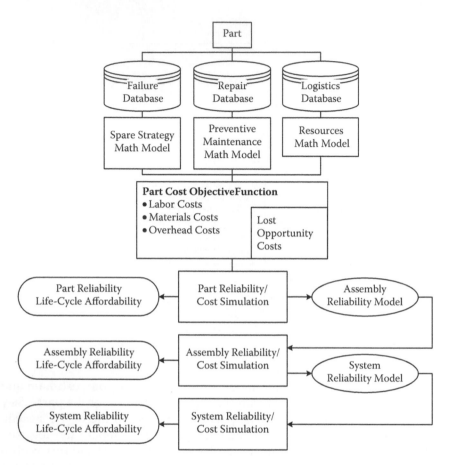

FIGURE 1.12
Reliability life-cycle reliability simulation logic.

Life-cycle reliability simulations for parts, assemblies, and systems assume that all repair and logistics resources are available at no cost. Reliability-based life-cycle economics analysis adds sustainment cost objective functions, part sparing strategies, and queuing models to the life-cycle reliability simulation model. Sustainment cost objective functions are developed for each part to include variable and fixed labor, materials, and overhead costs that apply to prerepair logistics downtime, repair time, and postrepair logistics downtime. All of the sustainment cost objective functions are included in the part life-cycle reliability models. Part sparing strategies include ad hoc spare parts acquisition, economic order quantity acquisition, and emergency acquisition, and are included in the part life-cycle reliability models. Queuing models characterize the wait time for logistics downtime events, repair time events, and postrepair logistics downtime events, and are included in the part life-cycle reliability models. The complete reliability and

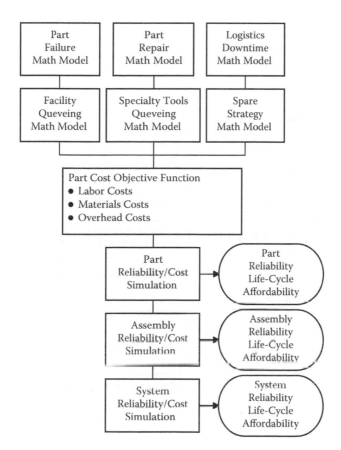

FIGURE 1.13
Reliability-based life-cycle economics analyses flowchart.

life-cycle economics simulation math model provides engineers and managers with the cost impact of part selection decisions and sustainment practices. Trade studies can be performed to improve the causes of all downtime from logistics downtime to repair time. The logic flow chart of reliability-based life-cycle economics analysis is provided in Figure 1.13.

Reliability-centered maintenance (RCM) is presented as a best practice for implementation of proactive maintenance. RCM enables engineers and managers to measure the risk-based consumed life of parts. RCM uses the reliability analysis performed for parts and the simulation tools to evaluate life-cycle economics analysis and to optimize the planning and performance of logistics and repair events at the lowest life-cycle costs. RCM follows three implementation paths: (1) condition-based maintenance (CBM), (2) time-directed maintenance (TDM), and (3) stress-directed maintenance (SDM). CBM measures a condition indicator that describes the consumed life of a part. TDM describes a risk-based time to consumed life. SDM measures the

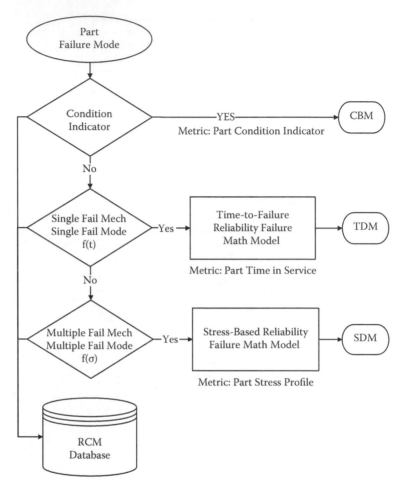

FIGURE 1.14
Reliability-centered maintenance decision logic.

stress profile for failure mechanisms acting on the part and the damage done to describe the risk-based consumed fatigue life of the part. The part failure math models, and the corresponding changes to the logistics and repair math models, are updated in the reliability-based life-cycle economics simulation to evaluate the economic impact from implementing RCM. The logic flow chart of RCM is provided in Figure 1.14.

A summary of the structure of the book is described below.

Chapter 2, Reliability Analysis for Part Design, addresses reliability fundamentals for part mission reliability, maintainability, and availability analyses that apply to part design.

Chapter 3, Reliability Analysis for System Sustainment, addresses reliability fundamentals for part mission reliability, maintainability, and availability analyses that apply to system sustainability.

[Implementation of reliability fundamentals for part mission reliability, maintainability, and availability analyses differ between design and sustainability applications, although the respective math models are identical.]

Chapter 4, Engineering Economic Analysis, addresses engineering economic fundamentals that apply to the evaluation of part life-cycle economic analysis.

Chapter 5, Reliability-Based Logistical Economic Analysis, addresses logistical resources allocation, cost estimation, and economic math models that apply to system life-cycle sustainability of parts applied to either design or sustainability.

Chapter 6, Life-Cycle Economic Analysis, incorporates reliability analysis, engineering economic analysis, and logistical economic math models to investigate life-cycle economic design and sustainability alternatives.

Chapter 7, Reliability-Centered Maintenance, addresses implementation of a reliability-centered maintenance program.

Chapter 8, Reliability Database, addresses requirements of reliability and economic database structure to efficiently acquire and use data from all of the reliability and economic analyses.

Chapter 9, Reliability Simulation and Analysis, addresses simulation and analysis methods that enable performance of reliability and economic analyses for parts, assemblies, and the system.

Detailed analytical approaches to fit reliability and economic models are provided in the appendices.

- Appendix A, Reliability Failure Math Models and Reliability Functions
- Appendix B, Maintainability Math Models and Maintainability Functions
- Appendix C, Logistics Downtime Functions
- Appendix D, Engineering Economics Functions

2

Reliability Analysis for Part Design

Reliability analysis in design provides the engineer and manager with an understanding of part failure.

Design for reliability is a process that begins with a part criticality analysis, continues with a qualitative part failure analysis for critical parts and a quantitative part failure analysis for critical parts, and concludes with reliability simulation for part, assembly, and system design configurations. Reliability analysis in design provides the information required to perform reliability-based life-cycle economic analysis.

- Criticality analysis is applied to all of the parts in the system and serves to reduce those parts to candidates for further reliability analysis. Criticality analysis culminates with a rank-ordered critical items list.

- Qualitative part failure analysis is applied to the parts contained in the critical items list. The qualitative part failure analysis serves to reduce the critical items list to the technically and economically feasible relevant few. It also provides the basis for empirical investigations performed in the quantitative failure analysis.

- Quantitative part failure analysis is performed to acquire empirical data from failure experiments and historical data to fit reliability math models.

- Part reliability simulations evaluate part selection alternatives.

- Assembly reliability simulations integrate part failure math models to the assembly design configuration to evaluate trade studies for modification to the design configuration, implementation of preventive maintenance practices, and evaluation of resource constraints.

- System reliability simulations are conducted on the critical parts for the entire system to evaluate conformance to the reliability requirements and evaluation of what-if analyses performed at the part and assembly levels.

The process is illustrated in the flowchart in Figure 2.1.

We carry out the design for reliability process in order to map the procedures for performing probabilistic risk assessments for

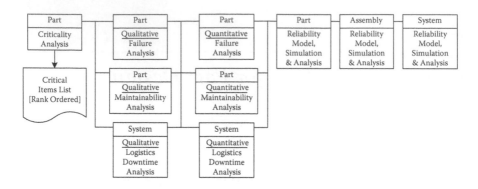

FIGURE 2.1
Design for reliability process map.

1. Technical constraints

   Risk that the part design integrated into the system design configuration will not meet the design organization's or end user's functional and reliability requirements.

2. Cost constraints

   Risk that the part design integrated into the system design configuration will not meet the economic requirements for the design organization or the end user.

3. Schedule constraints

   Risk that the performance of part design will not meet the system design and development plan for the design organization or meet the end user's expectation for operational availability.

Part Failure

In this section we will establish the lexicon of reliability analysis.

The most basic unit of a reliability analysis is a part. A part is defined as an item that is removed and replaced in an assembly to restore the system to an up state. The designation of a part goes by a variety of names, lowest replaceable unit, component, and item, to name a few. Too often design engineers will specify a part using the vendor nomenclature. Vendor nomenclature may refer to the part as an assembly, subsystem, or system. It is common to find a bill of materials for an assembly where vendor nomenclature and part numbers are included in the parts list.

To further muddy the waters, the maintenance practices performed by different organizations for the same part can skew the part designation as defined above. Consider the air brake canister for a mining haul truck.

One maintenance practice may treat the air brake canister as a part that is removed and replaced as a single unit. Another maintenance practice may treat the air brake canister as an assembly in which the internal component of the air brake canister is removed and replaced. In the former case the haul truck is restored for operations following the removal and replacement of the air brake canister, in the latter case the haul truck is restored for operations following the removal and replacement of the internal component of the air brake canister.

There are two kinds of parts: repairable and irreparable. Irreparable parts are removed and disposed of. Repairable parts are taken to a shop and rebuilt. The internal components used to rebuild a repairable part are frequently referred to as a shop repairable unit. The distinction between repairable and irreparable parts will be important in the qualitative and quantitative failure analysis tasks.

The reader should note what we mean by failure: systems do not fail, subsystems do not fail, and assemblies do not fail. Only parts fail. Part failure results in an assembly being in a down state, the assembly may result in the subsystem being in a down state, and the subsystem may result in the system being in a down state.

Part failure is the result of material damage. A simple part consists of a single material, i.e., a V-belt used to drive a pulley. More often, parts are comprised of different materials that are fastened and joined. The different materials and fasteners fail in different ways.

A system can be in one of three states: available, in use, and down. A system that is available is idle and fully capable of beginning the next mission. A system that is in use is operating at full or degraded design states. A system that is down is incapable of performing its mission until maintenance actions restore it to an available state. Figure 2.2 shows a basic timeline of the three states.

Part Criticality Items List and Database

The part criticality analysis recognizes that resource constraints, technical constraints, cost, and schedule prevent a reliability analysis for all parts in a system design. Part criticality analysis is the initial step to reduce the entire bill of materials to the technically and economically feasible relevant few parts that require reliability analysis. Criticality analysis recognizes that all part failures are not equal: the effects of a part failure on the system design, on the system mission, and on the organization that uses the system are all varied.

Understanding the system design configuration down to the assemblies is the essential first step in performing a criticality analysis for parts. The system design configuration is typically provided by systems engineers as a work breakdown structure. A fault tree analysis of the work breakdown

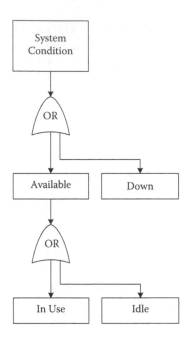

FIGURE 2.2
System condition flow chart.

structure provides an understanding of those parts that will begin the downing cycle from the assembly up to the system. The criticality analysis is applied to the parts with the use of a modified approach developed by John Moubray (1997). The first step in the criticality analysis is to determine the part failure consequences. The next step is to determine the perception of part failure to the operator and maintainer. The final step is to determine the interval between the perception of part failure and the failed state of the part. The results of the criticality analysis will be documented in a critical items list, and the information will be stored in a critical items database. The top level logic flow diagram for the criticality analysis is provided in Figure 2.3.

Design Configuration—Work Breakdown Structure

An illustrative work breakdown structure is provided in Figure 2.4 as an example of performing the part criticality analysis. The example system consists of subsystems that consist of assemblies that consist of parts.

Design Configuration by Fault Tree Analysis

The fundamental approach to perform a fault tree analysis on the work breakdown structure involves identifying logic gates that describe the conditions

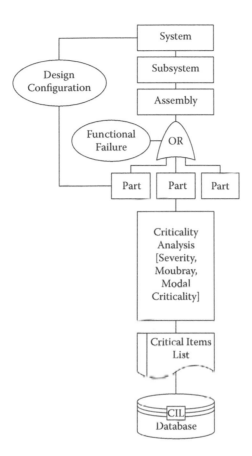

FIGURE 2.3
Critical items list logic.

that result in a down state due to a part failure. An example is provided in Figure 2.5. In this example, the system is in a down state if either subsystem is in a down state. The subsystem is in a down state if any of the assemblies are in a down state. The assembly is in a down state if any of the parts fail.

Modified Moubray Criticality Analysis

There are a variety of criticality analyses that have been used in design for several decades. They include the risk priority number, the modal criticality number, and a hazard or severity analysis. All of these methods require an understanding of the frequency and effects of part failure. That means that failure analysis must be performed on all parts in order to identify the critical items in the design. The objective here is to reduce the massive parts list down to the relevant few before investment is required to perform reliability analysis.

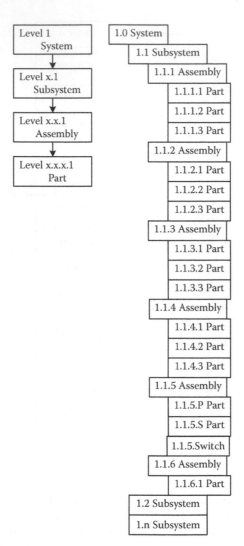

FIGURE 2.4
Work breakdown structure.

Consequences Analysis

The consequences analysis identifies the effect part failure will have on the system. There are four consequences:

- Catastrophic consequences describe part failure that results in system destruction and death or permanent injury to people. Catastrophic consequences can also include regulatory sanctions and prohibitive cost exposure.

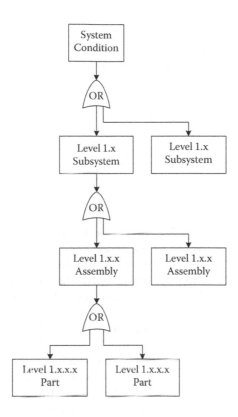

FIGURE 2.5
Fault tree analysis.

- Operational consequences describe part failure that results in a system downing event at the location that the part fails. Maintenance must be performed to restore the system to an up state before the system can be deployed for the next mission.
- Degraded mode consequences describe part failure that results in a design or mission degradation but does not result in a mission abort. The system is not in a down state, but maintenance must be performed at the conclusion of the mission.
- Run-to-failure consequences describe part failure that has no effect on the ability of the system to complete its current mission and to be deployed for its next mission. Typically the organization will allow this type of failure to occur because addressing the failure as it occurs is less costly than implementing a preventive maintenance practice.

Perception of Failure

The perception of failure describes the ability of the operator or the maintainer to perceive the failure mode of a part before the part fails. This perception can be

- Hidden to the operator during the mission and to the maintainer between missions
- Evident to the operator during the mission and to the maintainer between missions

P-F/P-M Interval

John Moubray describes the P-F interval as the time between the perception of failure and the occurrence of the failed state. The modified P-F interval, known as the P-M interval, is defined as the duration between the perception of failure and the initiation of preventive maintenance actions. The P-F interval is an on-condition indicator that provides no opportunity to plan a maintenance action to preserve the system's functionality. Immediate maintenance is required. The P-M interval is a condition indicator that measures the consumed life of the part that provides the opportunity to plan a maintenance action to preserve the system's functionality.

The P-F interval can be either greater than the mission duration or less than the mission duration. The operator incurs limited risk in completing a mission when the P-F interval is greater than the mission duration. The operator incurs the risk associated with aborting a mission in place when the P-F interval is less than the mission duration. The organization incurs no risk in performing a mission when the P-M interval is greater than the mission duration. A preventative maintenance program does not allow the P-M interval to be less than the mission duration, unlike the P-F interval, which can be less than the mission duration.

Criticality Analysis Rank

The modified Moubray criticality analysis can rank order the parts based first on the consequences of failure where catastrophic consequences rank highest, operational consequences rank second, and degraded mode consequences rank last. Criticality analysis is not performed on parts that have a run-to-failure consequence. Within each consequence, a hidden perception of the failure mode ranks higher than an evident perception. The hidden perception of failure does not have a P-F interval. Evident perceptions of failure have a P-F interval, and a P-F interval less than mission duration ranks higher than a P-F interval greater than mission duration. The ranking order is defined below.

I. Catastrophic

   1. Hidden

   2. Evident, P-F < mission duration

   3. Evident, P-F > mission duration

II. Operational

   1. Hidden

   2. Evident, P-F < mission duration

   3. Evident, P-F > mission duration

III. Degraded mode—design and mission

   1. Hidden

   2. Evident, P-F < mission duration

   3. Evident, P-F > mission duration

IV. Run to failure

Proposed Criticality Analysis Procedure

The flow diagram in Figure 2.6 describes the procedure for the modified Moubray criticality analysis. The procedure introduces a brainstorming and red team approach to performing this criticality analysis. The brainstorming approach uses an organization's employees who understand the part and the system to perform the criticality analysis. The brainstorming approach has a single rule—no judgments are made. The red team is comprised of an organization's more experienced employees. It is the red team that makes judgments about the findings of the brainstorming team. Then the red team reduces all of the brainstorm analysis to the meaningful critical items list.

Reliability Block Diagram

The parts contained in the critical items list are shown in the system design configuration down to the assemblies using fault tree logic in Figure 2.7.

Or-gates define a serial design configuration for the assembly and higher design levels. And-gates define a redundant design configuration for the assembly and higher design levels. The revised system design

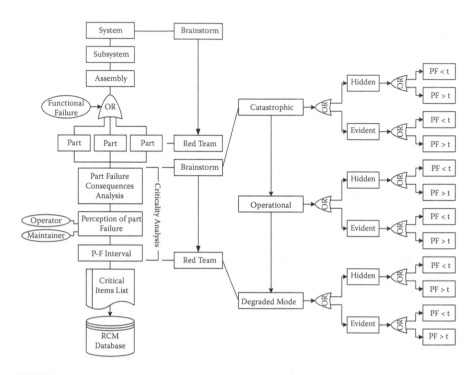

FIGURE 2.6
Criticality analysis flow chart.

configuration is expressed as a reliability block diagram illustrated in Figure 2.8.

The entire system is represented in the reliability block diagram by the critical parts only. The previous fault tree for the system defines the first subsystem as critical. The first subsystem defines six assemblies that are critical to the subsystem. The assemblies are a serial design configuration and are single points of failure for the subsystem. Assembly 1 is comprised of critical parts in a serial design configuration. Assembly 2 is comprised of a critical part in an active parallel redundant design configuration that has the risk of catastrophic or operational consequences. Assembly 3 is comprised of parts in a serial-in-parallel redundant design configuration. Loss of one serial path reduces assembly 3 to a serial design configuration with catastrophic or operational consequences. Assembly 4 is comprised of parts in a parallel-in-serial redundant design configuration. Loss of one part in a parallel path reduces assembly 4 to a serial design configuration with catastrophic or operational consequences. Assembly 5 is comprised of parts in a standby redundant design configuration. Loss of the primary part reduces assembly 5 to a serial design configuration with catastrophic or operational consequences. Assembly 6 is comprised of parts in an n-provided r-required

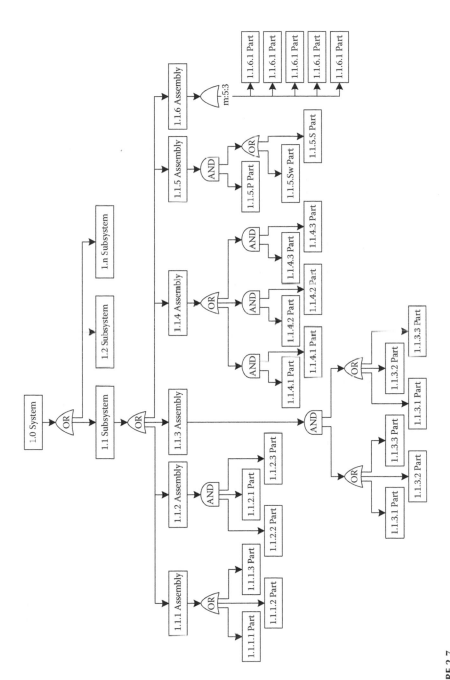

FIGURE 2.7
System work breakdown structure.

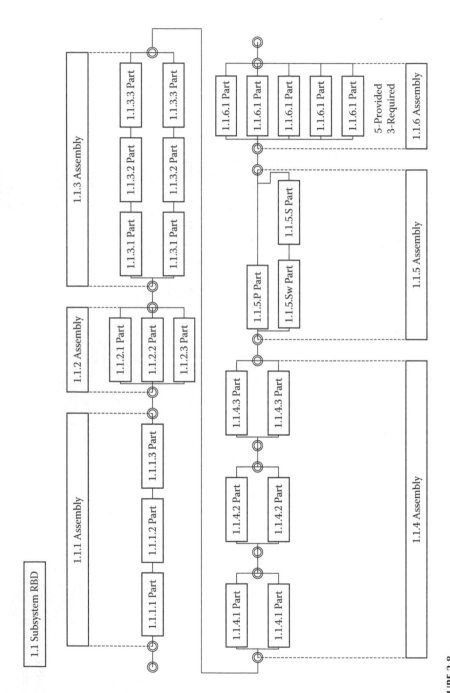

FIGURE 2.8
Reliability block diagram.

redundant design configuration. Loss of $n - r$ parts reduces assembly 6 to a serial design configuration with catastrophic or operational consequences.

The criticality analysis is complete when there are sufficient parts in the critical items list to begin the qualitative part failure analysis. It is unrealistic to delay the qualitative part failure analysis until the criticality analysis for the entire system work breakdown structure is completed.

Qualitative Part Failure Analysis

The qualitative part failure analysis consists of identifying the source of failure mechanisms acting on a critical part, the failure mechanisms acting on a critical part, the failure modes caused by the failure mechanisms, and the failure effect for each failure mode as presented in Figure 2.9.

Sources of failure mechanisms include the following:

- Transportation—all events associated with packing, material handling, and transporting the part throughout the supply chain
- Storage—all events associated with material handling and storage of the part throughout the supply chain
- Functional—operation of the part and the system in which it is installed (Functional sources of failure mechanisms are classified as operational [intrinsic] and ambient [extrinsic].)
- Operational—events resulting from the function of the part or stresses that result from part operation
- Ambient—events resulting from the function of proximate parts to the part or stresses that are applied to the part due to the operation of other parts

The sources of failure mechanisms are the conditions of use identified in the definition for reliability, "the probability that a part will function without failure for the specified mission duration under stated conditions of use."

Failure Mechanisms

Failure mechanisms are stresses that act on the part. Examples of failure mechanisms from all sources include:

- Corrosive agents
- Humidity
- Salt spray

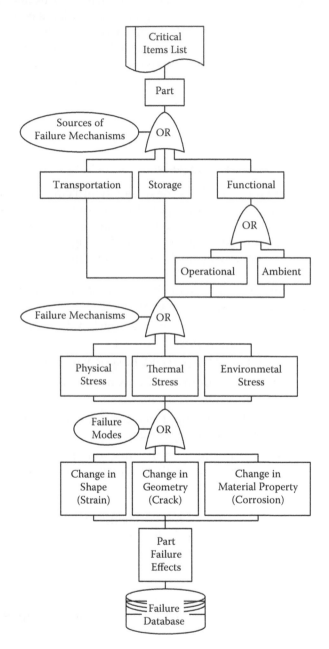

FIGURE 2.9
Qualitative part failure analysis.

- Sand exposure
- Thermal shock
- Vibration
- Physical shock
- Thermal loads
- Vibration
- Tension
- Compression
- Torsion

Failure Metrics

Identification of failure mechanisms that act on the part must also include the metrics that describe the magnitude of the failure mechanism. Failure metrics include measures of force, temperature, gravitational loading, etc.

Part failure is often the result of two or more failure mechanisms. Failure mechanisms can act independently of each other. Failure mechanisms can be dependent. The interaction between two or more failure mechanisms can cause failure in some cases where one failure mechanism acting alone would not. These relationships are illustrated in Figure 2.10. The figure

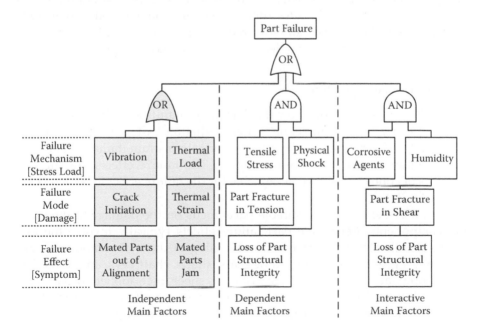

FIGURE 2.10
Failure mechanism investigation.

highlights the treatment of failure mechanisms as independent main factors because the dominant approach to reliability failure analysis makes that assumption. An investigation of failure mechanisms must include recognition of dependency and interactions.

Failure Modes

Failure modes are caused by failure mechanisms. Part failure modes are the damage to the part, the materials comprising the part, and interfaces between parts. Jack Collins observed that there are three categories of failure mechanisms for mechanical and structural parts:

- Change in shape—longitudinal and torsional strain
- Change in geometry—pitting, galling, crack initiation, fracture
- Change in material properties—corrosion, intermetallic compounds

Failure Mode Metrics

Change in shape can be measured by dimension analysis. Change in geometry can be measured by crack dimensions, number of test or call points, and the existence or dimension of cracks. Change in material properties is less straightforward. Visual evidence of corrosion reveals little about the actual condition of the part. Nondestructive examination methods, to include X-ray, acoustical resonance, and infrared scanning, can provide insights into changes in the material's mass density, elasticity, and strength.

Failure Effects—Symptom

Part failure effects describe how the damage is manifest by the design integrity of the part. For example, ultraviolet radiation (failure mechanism) acting on a V-belt driving a pulley causes elongation strain (failure mode) and delamination of the plies (failure mode). The failure effects are slipping and loss of contact. It is never acceptable for the failure effect to be described as "part fails."

Part failure effects for repairable parts focus on the damage to the internal components. Repairable part failure effects typically involve the loss of structural integrity of fastener and joining of internal components. For example, a gearbox is comprised of gears, shafts, shaft keys, and seals. The operational failure mechanism described as sheer force applied to gear teeth can result in multiple failure modes, chipped gear teeth, sheared shaft keys, etc. The failure effect for the gearbox is loss of applied torque.

Failure Effects Metrics

It is not always possible to define a metric for a failure effect. A change in assembly capacity is one example. The rate of change for moving parts may be subjectively characterized.

Failure Effects for an Assembly

Describing the failure effect for an assembly is analogous to the failure effect for a repairable part. Assembly failure effects are descriptions of the symptoms resulting from part failure modes and effects. Another issue with assembly failure effects is the design configuration of the parts in the assembly. An assembly comprised of parts in a serial design configuration will manifest failure effects that describe an assembly downing event. An assembly comprised of parts in a redundant design configuration will manifest a degraded mode of the design and the assembly's capacity. Consider the following scenarios:

1. Hard part failure
   - Serial assembly in down state
   - Parallel assembly in degraded design state visible to assembly condition metrics
2. Part wear out (fatigue)
   - Serial assembly in degraded capability state
   - Parallel assembly in degraded capability state not visible to assembly condition metrics
3. Soft part failure
   - Serial assembly experiences intermittent down state not requiring maintenance to resume up state
   - Parallel assembly state not visible to assembly condition metrics

The qualitative failure analysis is complete when there are sufficient parts from the critical items list to begin the quantitative part failure analysis.

Quantitative Failure Analysis

The quantitative failure analysis uses the information from the qualitative failure analysis and the critical items list to identify parts that require empirical investigation to fit reliability math models. Reliability math models include the following:

- Part failure math model
- Part cumulative failure math model
- Part survival function
- Part mission reliability model
- Part hazard function
- Part repair math model
- Part logistics downtime math model

Quantitative Reliability Analysis

Quantitative reliability analysis is the procedure for fitting empirical failure data to reliability math models. The part failure math model is the basis for determining whether the part, and by extension the system, meets or exceeds the reliability requirements for the system. The procedure is to determine the reliability allocation for the critical part, acquire empirical failure data from historical records or experiments, and fit the data to the appropriate reliability failure math model; either time to failure or the stress strength. The procedure is illustrated in Figure 2.11.

Part Reliability Allocation

The design engineer requires an estimate for the reliability requirement for a part. The allocation of the system reliability requirement is performed by systems engineering to provide that information. The most common best practice for part reliability allocation is known as the equal allocation method. The equal allocation method assumes that all subsystems are in serial design configuration, all assemblies in a subsystem are in serial design configuration, and all parts in an assembly are in a serial design configuration. The system reliability requirement is expressed as a percentage, i.e., 99%. The equal allocation method reverses the procedure for calculating the reliability of a design configuration that is comprised of two or more serial components.

$$R_j = \sqrt[n]{R_i} \tag{2.1}$$

For example, if an assembly is comprised of two parts that have a reliability of 90% and 95%, then the reliability of the next higher design configuration is equal to the product of the two part reliabilities, 90% × 95% = 85.5%. Extending the example, if the assembly is allocated a reliability of 85.5%,

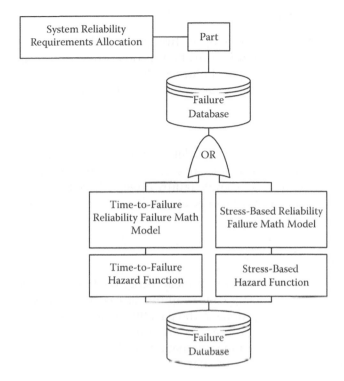

FIGURE 2.11
Quantitative failure analysis and failure database.

then the equal allocation of the part reliability will be equal to the value that would yield 85.5%, or the square root of 85.5% = 92.5%. The design engineers will treat the reliability allocation as a budget to evaluate the design decisions that they make from the design analysis they perform. The serial design configuration assumption is based on the fact that the dominant approach to converting a system functional requirement to lower-level design configurations yields a serial design. For example, consider a car. Reduce the total number of parts in the car to only those that are critical to the operation of the car. You may be surprised to find that the car is a serial design configuration. The introduction of redundancy in a design is typically driven by the inability to meet the reliability requirement with a serial design, the need to increase the safety of the system, or the need to be able to take one or more components out of operation for predictive or preventive maintenance during the useful life of the system. We can summarize the equal reliability allocation for any design level as being equal to the nth root of the reliability allocation of the next higher design level. If a subsystem is comprised of 12 assemblies, the equal reliability allocation of each assembly will be the 12th root of the reliability allocation of the subsystem.

Failure Math Models

Failure math models are developed through empirical investigation of part failure. The qualitative part failure analysis determines the design of experiments for the quantitative part failure analysis.

There are two approaches to quantitative part failure analysis:

Time to Failure

Time-to-failure reliability math models assume that a part fails in time. This is an acceptable assumption for parts that have a single failure mechanism, a single failure mode, and a single failure effect, where the failure mode correlates to time. Units of time can be chronological, engine hours, cycles, cold starts, etc.

Stress Strength

Stress-strength reliability math models recognize that time does not cause failure, failure mechanisms cause failure. Stress-strength reliability math models show the damage done to the part by stress loads.

Although reliability requirements are often stated in terms of reliability, the most important design and sustainability metric is the hazard function, the instantaneous failure rate of the part.

Reliability Functions

Quantitative part failure analysis fits math models that describe the reliability functions. The empirical data are acquired from historical data of part failure and failure experiments. Empirical data are stored in the part failure database. The reliability functions are provided in detail in Appendix A. The procedure to express the reliability functions includes the following:

1. Descriptive statistics and frequency distribution of sample data

   The empirical data are recorded in a spreadsheet. The descriptive statistics of the data are calculated to include the sample mean and standard deviation, the sample kurtosis and skewness, and the sample size. The data are fit to a frequency distribution that plots the sample histogram. The sample histogram describes the shape of the distribution of the data: exponential, skewed, or symmetrical about the mean.

2. Failure probability distribution, pdf, $f(t)$

   The data are fit to a probability mass density distribution, the pdf, expressed as $f(t)$, for time-to-failure data, or $f(x)$, for stress-strength data. The failure probability distribution describes the behavior of the failure mode over time or applied stress loads.

3. Cumulative failure probability distribution, cdf, $F(t)$

The cumulative probability mass density distribution, the cdf, expressed as $F(t)$, for time-to-failure data, or $F(x)$, for stress-strength data, is calculated as the integral of $f(t)$ or $f(x)$ from 0 to t or x, respectively. The cumulative failure probability distribution describes the cumulative proportion of failure over time or applied stress loads. The cumulative failure probability distribution is equal to 0 when t or x equal 0. The cumulative failure probability distribution approaches 1 as t or x approach infinity.

4. Survival function, $S(t)$

The survival function, expressed as $S(t)$, for time-to-failure data, or $S(x)$, for stress-strength data, is calculated as the complement of the cumulative probability distribution, $S(t) = 1 - F(t)$. The survival function is equal to 1 when t or x equal 0. The survival function approaches 0 as t or x approach infinity. The definite integral of the survival function is equal to the mean time between failure for a part, or the mean time between downing events for a system.

5. Mission reliability function, $R(\tau|t)$ given $\tau \equiv$ mission duration

The mission reliability function for time-to-failure data, expressed as $R(\tau|t)$, is the conditional probability that a part will perform failure free for the next mission duration given that it has survived to the current time. The mission reliability function for stress-strength data is calculated using interference theory.

6. Hazard function, $h(t)$

The hazard function is the instantaneous failure rate of the part over time. It is the ratio of the failure probability distribution and the survival function and is expressed as $h(t) = f(t)/S(t)$.

Assembly Reliability Functions

Assembly reliability functions are comprised of part reliability functions based on the design configuration of the assembly.

Serial Design Configuration

The serial design configuration for an assembly is comprised of parts that are single points of failure for the assembly. The assembly will be in a down state if part one fails, if part two fails, or if part three fails. This is reflected by the or-gate in the fault tree in Figure 2.12. The serial assembly design configuration is expressed in a reliability block diagram as three parts in a series.

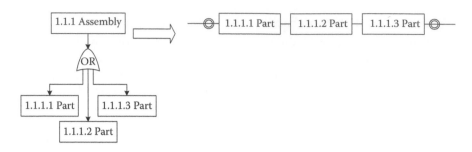

FIGURE 2.12
Serial design configuration.

The assembly survival function for a serial design is equal to the product of the survival functions for each part. The mission reliability function for a serial design is equal to the product of the mission reliability functions for each part. A characteristic of a serial design is that the assembly mission reliability will be less than the minimum part mission reliability.

$$S_{\text{Assembly}}(t) = \prod_{i=1}^{n} S_i(t) \tag{2.2}$$

$$R_{\text{Assembly}}(\tau \mid t) = \prod_{i=1}^{n} R_i(\tau \mid t) \tag{2.3}$$

$$R_{\text{Assembly}}(\tau \mid t) < \min\{R_i(\tau \mid t)\} \tag{2.4}$$

Active Parallel Design Configuration

The active parallel design configuration for an assembly is comprised of three parts that all function together. The assembly will be in a down state if all three parts fail. This is reflected by the and-gate in the fault tree in Figure 2.13. The active parallel assembly design configuration is expressed in a reliability block diagram as three parts in parallel.

The assembly survival function is equal to

$$1 - \prod_{i=1}^{n} (1 - S_i(t)).$$

Since 1 less the survival function is equal to the cumulative failure distribution, the assembly survival function can be restated as

$$1 - \prod_{i=1}^{n} (F_i(t)).$$

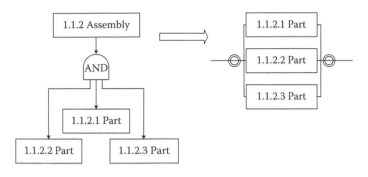

FIGURE 2.13
Active parallel design configuration.

The assembly reliability function is equal to

$$1 - \prod_{i=1}^{n} (1 - R_i(\tau|t)).$$

The unreliability of a part is equal to $1 - R_i$. It is a characteristic of active parallel redundant design configurations that the assembly reliability will be greater than the maximum part reliability in the design.

$$S_{\text{Assembly}}(t) = 1 - \prod_{i=1}^{n}(1 - S_i(t)) = 1 - \prod_{i=1}^{n} F_i(t) \qquad (2.5)$$

$$R_{\text{Assembly}}(\tau|t) = 1 - \prod_{i=1}^{n}(1 - R_i(\tau|t)) \qquad (2.6)$$

$$R_{\text{Assembly}}(\tau|t) > \max\{R_i(\tau|t)\} \qquad (2.7)$$

Serial-in-Parallel Design Configuration

The assembly serial-in-parallel design configuration (Figure 2.14) is a typical result of a serial design that is found to not meet the reliability requirements. The functional and interface characteristics of each of the parts in the serial design prevent the inclusion of any parallel design for an individual part in the assembly. The alternative design requires replication of the serial path. The inclusion of the parallel serial path is reflected by the and-gate for the parts in the added path. The reliability block diagram represents the serial paths in parallel.

Calculation of the assembly survival function is a two-step procedure. The first step is to calculate the survival function for each path, designated *j*.

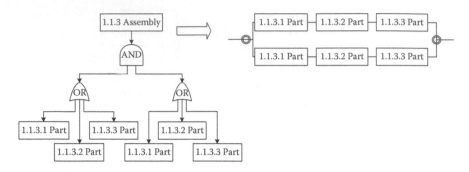

FIGURE 2.14
Serial-in-parallel design configuration.

The survival function of any path j is equal to the product of the survival functions for the parts in the serial design. The assembly survival function is then calculated by treating each path as a parallel part. In this example the assembly survival function is equal to

$$1-\prod_{j=1}^{m}\left(1-S_j(t)\right).$$

The assembly reliability function is calculated in the same two-step procedure. The reliability function for each path is calculated as the product of the reliability of the serial parts in the design (Figure 2.15). The assembly reliability function is then calculated as if each path was a parallel part (Figure 2.16). The assembly reliability function is equal to

$$1-\prod_{j=1}^{m}\left(1-R_j(\tau\,|\,t)\right).$$

$$S_j(t)=\prod_{i=1}^{n}S_i(t) \tag{2.8}$$

FIGURE 2.15
Serial path.

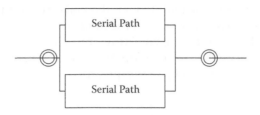

FIGURE 2.16
Parallel path.

$$R_j(\tau \mid t) = \prod_{i=1}^{n} R_i(\tau \mid t) \tag{2.9}$$

$$S_{\text{Assembly}}(t) = 1 - \prod_{j=1}^{m} \left(1 - S_j(t)\right) \tag{2.10}$$

$$R_{\text{Assy}}(\tau \mid t) = 1 - \prod_{j-1}^{m} \left(1 - R_j(\tau \mid t)\right) \tag{2.11}$$

Parallel-in-Serial Design Configuration

The assembly parallel-in-serial design configuration is also a typical result of a serial design that is found to not meet the reliability requirements. However, in this case, the functional and interface characteristics of each of the parts in the serial design do allow the inclusion of parallel design configurations for any one or all of the parts. In this case the output of any part in a pair can be an input to any part in the next pair. Each pair is in series. The reliability block diagram reflects the parallel design configuration (Figure 2.17)

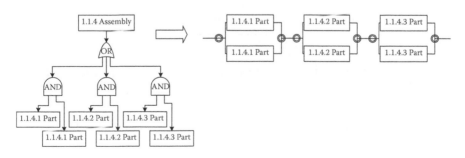

FIGURE 2.17
Parallel-in-serial design configuration.

of each pair as shown in the fault tree by the and-gate and the serial design configuration of the pairs as shown in the fault tree by the or-gate.

Calculation of the assembly survival function is a two-step procedure. The first step is to calculate the survival function for each parallel pair. The survival function for each parallel pair is equal to

$$1 - \prod_{i=1}^{n} \left(1 - S_i(t)\right).$$

Each pair now constitutes an equivalent part in series (Figure 2.18). The assembly survival function is equal to the product of the survival functions of each equivalent part. Again, the assembly mission reliability is calculated in the same manner that the assembly survival function is calculated.

$$S_{\text{Assembly}}(t) = \prod_{j=1}^{m} S_j(t) \tag{2.12}$$

$$R_{\text{Assembly}}(\tau \,|\, t) = \prod_{j=1}^{m} R_j(\tau \,|\, t) \tag{2.13}$$

$$S_j(t) = 1 - \prod_{i=1}^{n} \left(1 - S_i(t)\right) \tag{2.14}$$

$$R_j(\tau \,|\, t) = 1 - \prod_{i=1}^{n} \left(1 - R_i(\tau \,|\, t)\right) \tag{2.15}$$

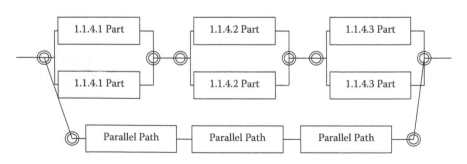

FIGURE 2.18
Equivalent serial path.

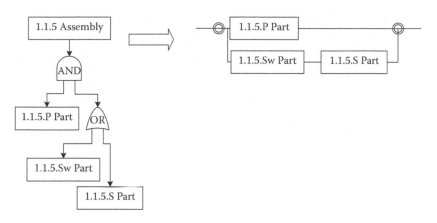

FIGURE 2.19
Standby design configuration.

Standby Design Configuration

The assembly standby design configuration is also a typical result of a serial design that is found to not meet the reliability requirements (Figure 2.19). However, in this case, the functional and interface characteristic of one of the parts in the serial design does allow the inclusion of a parallel design configuration. The option exists that the redundancy be an active parallel design configuration. Another option is to provide a standby design configuration where the secondary part is idle until the primary part fails. A switch mechanism activates the secondary part. The secondary part is operational until the primary part is replaced by a new part. The primary part is always the priority part. The assembly is in a down state if the primary and the secondary parts fail. The secondary part fails when the switch or the secondary part fails.

There are four alternatives for a standby design configuration:

1. Equal parts–perfect switch design configuration

   Equal parts–perfect switch design configurations are comprised of a primary part and a secondary part where the reliability of the secondary part is identical to the reliability of the primary part, typically the secondary is identical to the primary. Assumption of the perfect switch is based on an electronic-digital design that has a failure rate measured in failures per billion hours. The assembly survival function assumes that the primary and secondary parts have constant failure rates, λ, and are expressed as $e^{-\lambda t} \times (1 - \lambda \times t)$. The assembly reliability function is expressed as the survival function evaluated for the mission duration, τ.

$$S_{SB\_EP}(t) = e^{-\lambda_{Primary}t}\left(1-\lambda_{Primary}t\right) \tag{2.16}$$

$$R_{SB\_EP}(\tau) = e^{-\lambda_{Primary}\tau}\left(1-\lambda_{Primary}\tau\right) \tag{2.17}$$

2. Equal parts–imperfect switch design configuration

Equal parts–imperfect switch design configurations are comprised of a primary part and a secondary part where the reliability of the secondary part is identical to the reliability of the primary part. The switch has a nontrivial probability of failure, P_{Switch}. Assembly survival function assumes that the primary and secondary parts have constant failure rates, λ, and is expressed as $e^{-\lambda t} \times (1 - P_{Switch} \times \lambda \times t)$. The assembly reliability function is expressed as the survival function evaluated for the mission duration, tau, τ.

$$S_{SB\_EI}(t) = e^{-\lambda_{Primary}t} \left(1 - P_{Switch}\lambda_{Primary}t\right) \qquad (2.18)$$

$$R_{SB\_EI}(\tau) = e^{-\lambda_{Primary}\tau} \left(1 - P_{Switch}\lambda_{Primary}\tau\right) \qquad (2.19)$$

3. Unequal parts–perfect switch design configuration

Unequal parts–perfect switch design configurations are comprised of a primary part and a secondary part where the reliability of the secondary part is less than the reliability of the primary part. The switch is assumed to be perfect as in the previous example. Assembly survival function assumes that the primary and secondary parts each have constant failure rates, $\lambda_{Primary}$, $\lambda_{Standby}$, respectively, and is expressed as $e^{-\lambda_{Primary}t} + ((\lambda_{Primary}/\lambda_{Standby} - \lambda_{Primary}) \times (e^{-\lambda_{Primary}t} - e^{-\lambda_{Standby}t}))$. The assembly reliability function is expressed as the survival function evaluated for the mission duration, τ.

$$S_{SB\_UP}(t) = e^{-\lambda_{Primary}t} + \frac{\lambda_{Primary}}{\lambda_{Standby} - \lambda_{Primary}}\left(e^{-\lambda_{Primary}t} - e^{-\lambda_{Standby}t}\right) \qquad (2.20)$$

$$R_{SB\_UP}(\tau) = e^{-\lambda_{Primary}\tau} + \frac{\lambda_{Primary}}{\lambda_{Standby} - \lambda_{Primary}}\left(e^{-\lambda_{Primary}\tau} - e^{-\lambda_{Standby}\tau}\right) \qquad (2.21)$$

4. Unequal parts–imperfect switch design configuration

Unequal parts–perfect switch design configurations are comprised of a primary part and a secondary part where the reliability of the secondary part is less than the reliability of the primary part. The switch has a nontrivial probability of failure, P_{Switch}, as in the previous example. Assembly survival function assumes that the primary and secondary parts each have constant failure rates, $\lambda_{Primary}$, $\lambda_{Standby}$, respectively, and is expressed as $e^{-\lambda_{Primary}t} + ((\lambda_{Primary}/\lambda_{Standby} - \lambda_{Primary}) \times P_{Switch} \times (e^{-\lambda_{Primary}t} - e^{-\lambda_{Standby}t}))$. The assembly reliability function is expressed as the survival function evaluated for the mission duration, τ.

$$S_{SB\_UP}(t) = e^{-\lambda_{Primary}t} + \frac{\lambda_{Primary}}{\lambda_{Standby} - \lambda_{Primary}} P_{Switch}\left(e^{-\lambda_{Primary}t} - e^{-\lambda_{Standby}t}\right) \quad (2.22)$$

$$R_{SB\_UP}(\tau) = e^{-\lambda_{Primary}\tau} + \frac{\lambda_{Primary}}{\lambda_{Standby} - \lambda_{Primary}} P_{Switch}\left(e^{-\lambda_{Primary}\tau} - e^{-\lambda_{Standby}\tau}\right) \quad (2.23)$$

Problems with Current Standby Math Models

There are two problems with the current standby math models that are fortunately solved using simulation methods rather than algorithms.

1. The algorithms require fitting the failure math models to the exponential probability distribution. As we will see, there are other distributions that allow us to fit the failure data better than the exponential distribution. Therefore the constant failure rate used in these algorithms is no longer applicable.

2. The algorithms do not address the fact that the secondary part only ages when it is in use. So long as the primary part functions the secondary part is idle. After hundreds of hours the primary part may have gone through several failures and replacements. Meanwhile the standby part has only aged a few hours. This is illustrated in Figure 2.20.

n-Provided, *r*-Required Design Configuration

The *n*-provided *r*-required design configuration (Figure 2.21) is a special case of redundant design configurations. Consider a system that has several

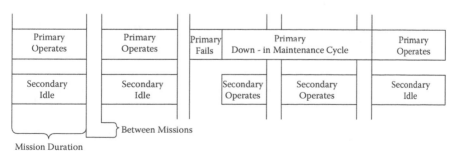

FIGURE 2.20
Standby logic.

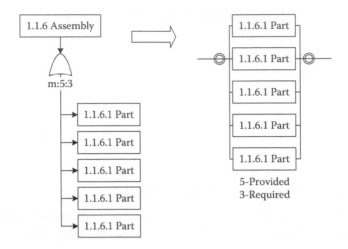

FIGURE 2.21
n-provided, *r*-required design configuration.

active parallel paths. The system is scheduled to operate continuously. The design engineers include an extra part in anticipation that maintenance will be performed while the system continues to operate. The assembly is in a down state when $r + 1$ parts fail.

There are two typical approaches to this design configuration:

1. *n* Provided, 1*n* Required

   This configuration is applicable when the cost of the part, or a corresponding path, is either too expensive or too large to accommodate more than one redundant part in the design configuration. The design engineers assume that the risk is acceptable to perform maintenance on one part either when the part fails or in a preventive maintenance practice.

2. *n* Provided, 2*n* Required

   This configuration is applicable when cost and size is not a consideration. The design engineers allow for one extra part to operate while one part is in maintenance.

The assembly survival function is the inverse cumulative binomial probability distribution in which the part survival function is the probability of success and one minus the part survival function is the probability of failure. The binomial probability distribution is calculated for *r* through *n* items. The mission reliability function is the inverse cumulative binary probability distribution in which the part reliability function is the probability of success and 1 less the part reliability function is the probability of failure.

$$S_{\text{Assembly}}(t) = \sum_{i=r}^{n}\left[\left(\frac{n!}{i!(n-i)!}\right)\right]\left(S_{\text{Part}}(t)\right)^{i}\left(1-S_{\text{Part}}(t)\right)^{n-i} \qquad (2.24)$$

$$R_{\text{Assembly}}(\tau\,|\,t) = \sum_{i=r}^{n}\left[\left(\frac{n!}{i!(n-i)!}\right)\right]\left(R_{\text{Part}}(\tau\,|\,t)\right)^{i}\left(1-R_{\text{Part}}(\tau\,|\,t)\right)^{n-i} \qquad (2.25)$$

The objective for presenting the survival and mission reliability functions is not to encourage the engineer to characterize assembly survival and reliability functions in their analysis. The number of parts and the various design configurations would render such an approach intractable. Rather the objective is to provide the engineer with an understanding of the survival and mission reliability functions that are used in commercially available reliability simulation software programs.

Fitting Reliability Data to Reliability Math Models

Commercially available reliability simulation software programs require that the engineer identify and fit failure, repair, and logistical downtime math models from empirical data. Two reliability failure math models are presented: the exponential and the Weibull distributions. Two reliability repair math models are presented: the log-normal and the Weibull distributions. One repair logistics downtime math model is presented—the triangular distribution. The methods used to fit the math models are presented in detail in Appendix A.

Exponential Reliability Math Models

The exponential reliability math model is fit by the arithmetic mean of the time-to-failure data, expressed in chronological time, hours, cold starts, etc. (Table 2.1). In this example, time is measured as cold starts not unlike the starter motor of an automobile. The mean time between failure is twenty missions. The failure rate is the inverse of the mean time between failure, 0.05 failures permission. The mission duration is one start per mission.

Table 2.2 describes how the exponential failure math model behaves. Assume that 100 systems, N_0, are available at the beginning of their useful life. At the conclusion of the first mission, five parts will fail ($r = \lambda N_0$), $f(t)$.

TABLE 2.1

Exponential Reliability Failure
Math Model

θ (missions)	20
λ (fpm)	0.05
τ (missions)	1

TABLE 2.2

Exponential Reliability Failure Math Model Data

m	N	r	$F(m)$	$S(m)$	$R(m)$
0	100	0	0	100	0.95
1	95.00	5.00	5.00	95	0.95
2	90.25	4.75	9.75	90.25	0.95
3	85.74	4.51	14.26	85.74	0.95
4	81.45	4.29	18.55	81.45	0.95
5	77.38	4.07	22.62	77.38	0.95
6	73.51	3.87	26.49	73.51	0.95
7	69.83	3.68	30.17	69.83	0.95
8	66.34	3.49	33.66	66.34	0.95
9	63.02	3.32	36.98	63.02	0.95
10	59.87	3.15	40.13	59.87	0.95
11	56.88	2.99	43.12	56.88	0.95
12	54.04	2.84	45.96	54.04	0.95
13	51.33	2.70	48.67	51.33	0.95
14	48.77	2.57	51.23	48.77	0.95
15	46.33	2.44	53.67	46.33	0.95
16	44.01	2.32	55.99	44.01	0.95
17	41.81	2.20	58.19	41.81	0.95
18	39.72	2.09	60.28	39.72	0.95
19	37.74	1.99	62.26	37.74	0.95
20	35.85	1.89	64.15	35.85	0.95
21	34.06	1.79	65.94	34.06	0.95
22	32.35	1.70	67.65	32.35	0.95
23	30.74	1.62	69.26	30.74	0.95
24	29.20	1.54	70.80	29.20	0.95
25	27.74	1.46	72.26	27.74	

The cumulative failure distribution, $F(t)$, is equal to the five failed parts, $F(t) = \Sigma f(t)$. The survival function, $S(t)$, is equal to the 95 parts that have not failed $(1 - F(t))$. The mission reliability is equal to the survival function at the conclusion of the first mission divided by the survival function at the beginning of the useful life of the system, $S(t_1)/S(t_0)$. At the conclusion of the second mission, 4.75 parts will fail $(r = \lambda N_1)$. The cumulative failure distribution,

FIGURE 2.22
Exponential sample frequency distribution histogram.

$F(t)$, is equal to the 9.75 failed parts, $F(t) = \Sigma f(t)$. The survival function, $S(t)$, is equal to the 90.25 parts that have not failed $(1 - F(t))$. The mission reliability is equal to the survival function at the conclusion of the first mission divided by the survival function at the beginning of the useful life of the system, $S(t_2)/S(t_1)$. Observe that the mission reliability function remains constant for all missions. Also note in Table 2.2 that sixty-four parts have failed by the time the system reaches the mean time between failure of twenty missions.

The frequency distribution for the sample data is presented in Figure 2.22. It represents the expected shape of an exponential distribution.

The cumulative failure distribution and survival function is presented in Figure 2.23. Observe that they intersect at the median time–time to fail. Fifty percent of the parts will have failed by the conclusion of the 14th mission.

Exponential Part Failure Math Model

The exponential part failure math model, $f(t)$, for the data is fit by the mean time between failure, θ, as shown in the following equation. The exponential part failure math model is typically expressed in the alternate form using the failure rate, λ, rather than the mean time between failure.

$$f(t) = \frac{1}{\theta} e^{-\left(\frac{t}{\theta}\right)} \qquad (2.26)$$

$$\lambda = \frac{1}{\theta} \qquad (2.27)$$

$$f(t) = \lambda e^{-\lambda t} \qquad (2.28)$$

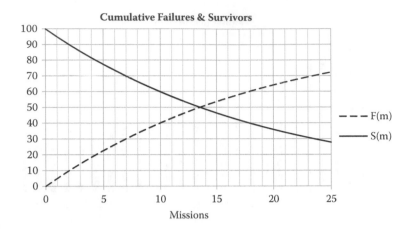

FIGURE 2.23
Exponential sample cumulative failure distribution and survival function.

Exponential Cumulative Part Failure Math Model

The exponential cumulative part failure math model, $F(t)$, is found by the integral of the part failure math model, as shown in the following equation.

$$F(t) = 1 - e^{-\lambda t} \tag{2.29}$$

Exponential Part Survival Function

The exponential survival function, $S(t)$, is the complement of the cumulative part failure math model, $S(t) = 1 - F(t)$, as shown in the following equation.

$$S(t) = e^{-\lambda t} \tag{2.30}$$

Exponential Part Mission Reliability Function

The exponential part mission reliability function is the conditional probability that the part will survive the next mission duration, τ, given that it has survived to time t. It is shown that the exponential part mission reliability function is constant for specified values for the mission duration since the failure rate and mission duration are constant.

$$R(\tau \mid t) = \frac{S(t+\tau)}{S(t)} = \frac{e^{-\lambda(t+\tau)}}{e^{-\lambda t}} = \frac{e^{-\lambda t}e^{-\lambda \tau}}{e^{-\lambda t}} = e^{-\lambda \tau} \tag{2.31}$$

$$R(\tau \mid t) = S(\tau) \tag{2.32}$$

In many texts and papers, the survival function is expressed as $R(t)$ and called the life function. This is because the exponential mission reliability function is equal to the survival function evaluated for the mission duration.

A part will have a constant reliability until the mission in which it fails, according to the exponential mission reliability function.

Exponential Part Hazard Function

The exponential part hazard function is its constant failure rate. The failure rate applies equally to a part on its first mission, the tenth, the hundredth, ad infinitum.

$$h(t) = \frac{f(t)}{1 - F(t)} = \frac{f(t)}{S(t)} = \frac{\lambda e^{-\lambda t}}{e^{-\lambda t}} = \lambda \tag{2.33}$$

Weibull Reliability Math Models

The following example for time-to-failure data presents a scenario that cannot be fit by the exponential failure math model but can be fit to the Weibull failure math model (Table 2.3). The time units are cold starts per mission, just like the exponential example. Assume that 100 systems, N_0, are available at the beginning of their useful lives. Notice that the incidence of part failures begins at the 17th mission. Until the 17th mission the part failure math model, $f(t)$, is 0, the cumulative failure math model is 0, the survival function is 1.0, and the mission reliability is 1.0.

At the conclusion of the 17th mission, two parts will fail, $f(t)$. The cumulative failure distribution, $F(t)$, is equal to the two failed parts, $F(t) = \Sigma f(t)$. The survival function, $S(t)$, is equal to the ninety-eight parts that have not failed $(1 - F(t))$. The mission reliability, $R(m)$ is equal to the survival function at the conclusion of the 17th mission divided by the survival function at the 16th mission, $R(m_{t17}) = S(t_{17})/S(t_{16})$. The hazard function, $h(m)$, is equal to the number of failures that occur on the 17th mission divided by the number of survivors at the completion of the 16th mission, $h(m_{t17}) = (r_{17}/S_{16})$. At the conclusion of the 18th mission, nine parts fail. The cumulative failure distribution, $F(t)$, is equal to the eleven failed parts, $F(t) = \Sigma f(t)$. The survival function, S_{16}, is equal to the eighty-nine parts that have not failed $(1 - F_{16}$, equals $100 - 11)$. The mission reliability is equal to the survival function at the conclusion of the 18th mission divided by the survival function 17th mission, $R(m_{t18}) = S(t_{18})/S(t_{17})$. Observe that the mission reliability function declines following the incidence of the first failure. Also note that the hazard function increases following the incidence of the first failure.

TABLE 2.3

Weibull Reliability Failure Math Model Data

m	N	r	F(m)	S(m)	R(m)	h(m)
0	100	0	0	100		0
1	100	0	0	100	1	0
2	100	0	0	100	1	0
3	100	0	0	100	1	0
4	100	0	0	100	1	0
5	100	0	0	100	1	0
6	100	0	0	100	1	0
7	100	0	0	100	1	0
8	100	0	0	100	1	0
9	100	0	0	100	1	0
10	100	0	0	100	1	0
11	100	0	0	100	1	0
12	100	0	0	100	1	0
13	100	0	0	100	1	0
14	100	0	0	100	1	0
15	100	0	0	100	1	0
16	100	0	0	100	1	0
17	98	2	2	98	0.980	0.020
18	89	9	11	89	0.908	0.092
19	71	18	29	71	0.798	0.202
20	38	33	62	38	0.535	0.465
21	8	30	92	8	0.211	0.789
22	0	8	100	0	0	1
23	0	0	100	0		
24	0	0	100	0		
25	0	0	100	0		

The frequency distribution for the sample data is presented in Figure 2.24. It represents a negatively skewed frequency distribution. The lack of part failures before the 17th mission suggests that a location parameter exists.

The cumulative failure distribution and survival function is presented in Figure 2.25. Observe that they intersect at the median time–time to fail. Fifty percent of the parts will fail by the conclusion of the 20th mission.

The mission reliability function is presented in Figure 2.26. Observe that the mission reliability is 1 until the 17th mission and then decreases over time.

The hazard function is presented in Figure 2.27. Observe that the hazard function is 0 until the 17th mission and then increases over time.

FIGURE 2.24
Weibull sample frequency distribution histogram.

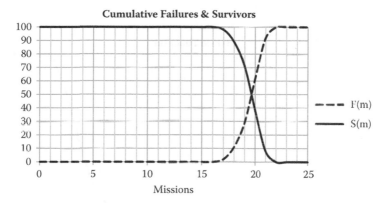

FIGURE 2.25
Weibull sample cumulative failure distribution and survival function.

Weibull Part Failure Math Model

The Weibull part failure math model for the data is fit from the failure data by the median ranks regression and coefficient correlation analysis. The Weibull failure math model is expressed in the following equation:

$$f(t) = \begin{vmatrix} 0 & \text{for} & t < \gamma \\ \left(\dfrac{\beta}{\eta^{\beta}}\right)(t-\gamma)^{\beta-1} e^{-\left(\frac{t-\gamma}{\eta}\right)^{\beta}} & \text{for} & t \geq \lambda \end{vmatrix} \qquad (2.34)$$

The following examples show the effects of the shape parameter, β, and the location parameter, γ, on the failure math model. Both Weibull failure

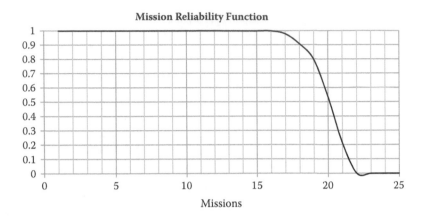

FIGURE 2.26
Weibull mission reliability function.

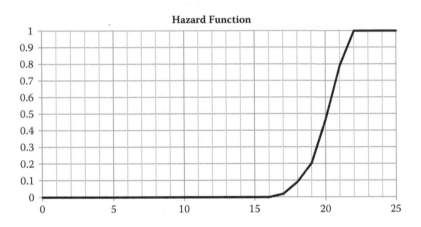

FIGURE 2.27
Weibull hazard function from sample data.

math models, $f_1(t)$ and $f_2(t)$, have equal values for the scale parameter, η. The first shape parameter, β_1, is 2.25 with a location parameter, γ, equal to 0. The Weibull failure math model for β_1 and $\gamma = 0$, is the positively skewed solid black line, $f_1(t)$. The Weibull failure math model for β_2 and $\gamma = 25$, is the negatively skewed dashed black line, $f_2(t)$ (Figure 2.28).

Weibull Cumulative Part Failure Math Model

The Weibull cumulative part failure math model is found by the integral of the part failure math model, as shown in the following equation.

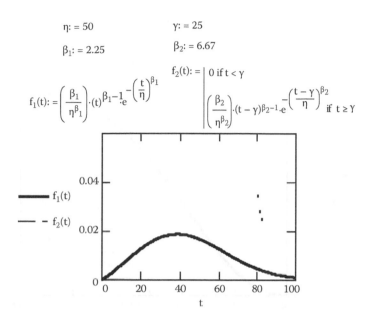

$\eta: = 50 \qquad\qquad \gamma: = 25$

$\beta_1: = 2.25 \qquad\qquad \beta_2: = 6.67$

$$f_1(t): = \left(\frac{\beta_1}{\eta^{\beta_1}}\right) \cdot (t)^{\beta_1-1} \cdot e^{-\left(\frac{t}{\eta}\right)^{\beta_1}}$$

$$f_2(t): = \begin{vmatrix} 0 & \text{if } t < \gamma \\[2ex] \left(\frac{\beta_2}{\eta^{\beta_2}}\right)\cdot(t-\gamma)^{\beta_2-1}\cdot e^{-\left(\frac{t-\gamma}{\eta}\right)^{\beta_2}} & \text{if } t \geq \gamma \end{vmatrix}$$

FIGURE 2.28
Weibull failure math model—shape parameter.

$$F(t) = \begin{vmatrix} 0 & \text{for} & t < \gamma \\[2ex] 1 - e^{-\left(\frac{t-\gamma}{\eta}\right)^{\beta}} & \text{for} & t \geq \lambda \end{vmatrix} \qquad (2.35)$$

The following examples show the effects of the shape parameter, β, and the location parameter, γ, on the cumulative failure math model (Figure 2.29). The Weibull cumulative failure math model for β_1 and $\gamma = 0$, is the solid black line, $F_1(t)$, which originates at the origin and approaches 1. The Weibull failure math model for β_2 and $\gamma = 25$, is the dashed black line, $F_2(t)$, which originates at the location parameter and approaches 1.

Weibull Part Survival Function

The Weibull survival function is the complement of the cumulative failure math model, as shown in the following equation.

$$S(t) = \begin{vmatrix} 1 & \text{for} & t < \gamma \\[2ex] e^{-\left(\frac{t-\gamma}{\eta}\right)^{\beta}} & \text{for} & t \geq \lambda \end{vmatrix} \qquad (2.36)$$

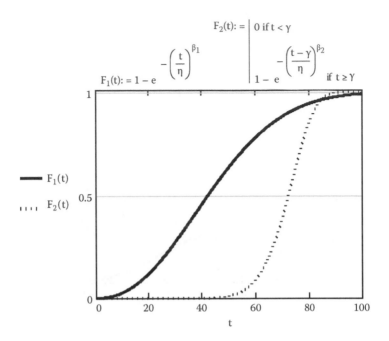

$$F_2(t):= \begin{vmatrix} 0 \text{ if } t < \gamma \\[2ex] 1-e^{-\left(\frac{t-\gamma}{\eta}\right)^{\beta_2}} & \text{if } t \geq \gamma \end{vmatrix}$$

$$F_1(t):= 1-e^{-\left(\frac{t}{\eta}\right)^{\beta_1}}$$

FIGURE 2.29

Weibull failure math model—location parameter.

The following examples show the effects of the shape parameter, β, and the location parameter, γ, on the survival function (Figure 2.30). The Weibull survival function for β_1 and $\gamma = 0$, is the solid black line, $S_1(t)$, which originates at 1 and approaches 0. The Weibull failure math model for β_2 and $\gamma = 25$, is the dashed black line, $S_2(t)$, which originates at the location parameter and approaches 0.

Weibull Part Mission Reliability Function

The Weibull mission reliability function is the conditional probability for the survival function evaluated for the mission duration given that the part has survived to time t, as shown in the following equation.

$$R(t) = \begin{vmatrix} 1 & \text{for} & t < \gamma \\[2ex] \dfrac{e^{-\left(\frac{t-\gamma+\tau}{\eta}\right)^{\beta}}}{e^{-\left(\frac{t-\gamma}{\eta}\right)^{\beta}}} & \text{for} & t \geq \lambda \end{vmatrix} \qquad (2.37)$$

The following examples show the effects of the shape parameter, β, and the location parameter, γ, on the mission reliability function (Figure 2.31). The

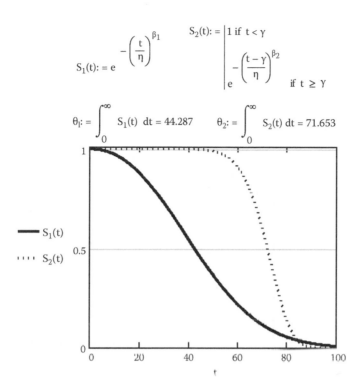

$$S_1(t) := e^{-\left(\frac{t}{\eta}\right)^{\beta_1}}$$

$$S_2(t) := \begin{vmatrix} 1 \text{ if } t < \gamma \\ e^{-\left(\frac{t-\gamma}{\eta}\right)^{\beta_2}} & \text{if } t \geq \gamma \end{vmatrix}$$

$$\theta_1 := \int_0^\infty S_1(t)\ dt = 44.287 \qquad \theta_2 := \int_0^\infty S_2(t)\ dt = 71.653$$

— $S_1(t)$

···· $S_2(t)$

FIGURE 2.30
Weibull survival function and MTBF.

Weibull mission reliability function for β_1 and $\gamma = 0$, is the solid black line, $R_1(\tau|t)$, which originates at 1 and approaches 0. The Weibull mission reliability for β_2 and $\gamma = 25$, is the dashed black line, $R_2(\tau|t)$, which originates at the location parameter and approaches 0.

Weibull Part Hazard Function

The Weibull hazard function is the ratio of the Weibull failure math model, $f(t)$, and the survival function, $S(t)$, as shown in the following equation.

$$h(t) = \begin{vmatrix} 0 & \text{for} & t < \gamma \\ \left(\frac{\beta}{\eta^\beta}\right)(t-\gamma)^{\beta-1} & \text{for} & t \geq \lambda \end{vmatrix} \qquad (2.38)$$

The following examples show the effects of the shape parameter, β, and the location parameter, γ, on the hazard function (Figure 2.32). The Weibull hazard function for β_1 and $\gamma = 0$, is the solid black line, $h_1(t)$, which originates at

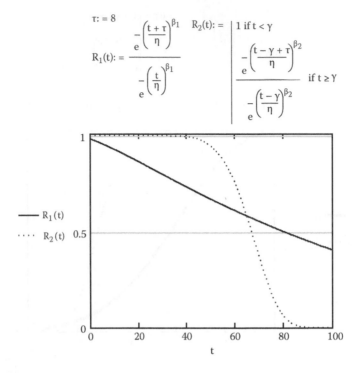

FIGURE 2.31
Weibull mission reliability function.

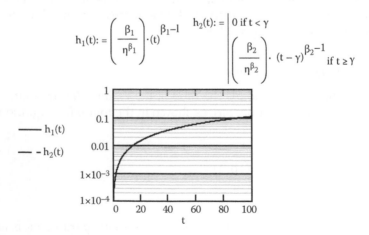

FIGURE 2.32
Weibull hazard function.

0 and approaches 1. The Weibull hazard for β_2 and $\gamma = 25$ is the dashed black line, $h_2(t)$, which originates at the location parameter and approaches 1.

The hazard function is best plotted on logarithmic-linear (natural log, Naperian) scales due to the very small magnitude for $h(t)$.

Qualitative Maintainability Analysis

Qualitative part maintainability analysis uses part damage and effects information to describe the repair procedures that must be understood to develop an effective maintainability practice.

Maintainability is the probability that a system will be restored to full functionality. Just as parts fail parts are repaired. Maintainability metrics are statistics developed from maintainability math models. Repair events are documented as work instructions. Repair experiments require the following:

- Awareness of resource requirements, to include specialty tools, maintenance manuals, fault diagnostic equipment, etc.
- Spare parts and installation hardware
- Labor skills and staffing levels
- Calibration instructions
- Repair verification procedures
- Documentation for failure accounting

The information acquired for each repair experiment is entered in the part repair database.

Efficacy of repair experiments by design engineers is often limited by several factors:

- The repair item is a prototype rather than a production unit that is delivered to the customer.
- The repair labor skills are prototype technicians rather than field mechanics.
- The repair environment is rarely comparable to the user's environment.

Quantitative Maintainability Analysis

The basic building block for quantitative maintainability analysis is the part repair math model. The part repair math model is fit from empirical time-to-repair data that is acquired from historical records and repair experiments. Time to repair is the sum of the measures of the durations to perform maintenance repair events. Time-to-repair data are typically positively skewed. The log-normal distribution has been the long standing best practice for fitting time-to-repair data to a repair math model. However, just as the Weibull distribution can fit skewed time-to-failure data, so too can it fit time-to-repair data.

Maintainability Metrics

The basic design maintainability metric is the mean time to repair (MTTR). The MTTR is found for the repair math model. The upper confidence limit for time to repair is calculated to understand the behavior of time to repair.

Log-Normal Approach

Time-to-repair data are measured in hours. But repair events often occur in minutes. Fitting the repair math model in minutes often provides better accuracy for mean time to repair, and the upper confidence limit that can be converted to hours.

The procedure for fitting TTR data to the log-normal distribution is as follows:

1. Record the TTR data in Excel, as shown in Table 2.4.
2. Calculate the descriptive statistics, to include the sample mean and standard deviation, the kurtosis and skewness, and the sample size, as shown in Table 2.5.

TABLE 2.4

Time-to-Repair Data in Minutes

121	132	147
127	134	156
139	116	149
147	140	114
124	161	124
156	139	137
128	109	121
119	126	145
116	113	111
171	126	130
132	134	131
183	163	191
117	165	172

TABLE 2.5

Time-to-Repair Descriptive Statistics

X_{bar} (min)	137.59
s (min)	20.63
Kurtosis	0.054
Skewness	0.833
n	39
X_{min} (min)	109
X_{max} (min)	191

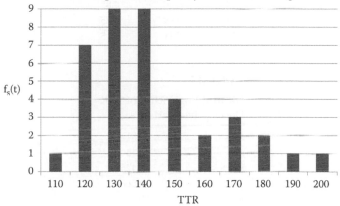

TTR Sample Data Frequency Distribution Histogram

FIGURE 2.33
TTR frequency distribution.

3. Plot the sample TTR frequency distribution and histogram, as shown in Figure 2.33.

   The histogram illustrates the positive skew of the TTR data.

4. Calculate the natural logarithm of the TTR data, as shown in Table 2.6.

5. Calculate the descriptive statistics for the logarithmic TTR data, to include the mean and standard deviation, and the sample size, as shown in Table 2.7.

6. Calculate the MTTR in minutes and convert to hours, as shown in Table 2.8. The MTTR is the anti-log of the mean of the natural logarithm of time-to-repair data:

$$\mu_{TTR} = e^{\mu_{LnTTR}} = e^{4.91} = 136.17 \tag{2.39}$$

TABLE 2.6

Natural Logarithm of TTR Data

4.80	4.88	4.99
4.84	4.90	5.05
4.93	4.75	5.00
4.99	4.94	4.74
4.82	5.08	4.82
5.05	4.93	4.92
4.85	4.69	4.80
4.78	4.84	4.98
4.75	4.73	4.71
5.14	4.84	4.87
4.88	4.90	4.88
5.21	5.09	5.25
4.76	5.11	5.15

TABLE 2.7

Natural Logarithm of TTR
Descriptive Statistics

X_{bar}	4.91
s	0.14
Kurtosis	−0.425
Skewness	0.566
n	39
X_{min}	4.69
X_{max}	5.25

TABLE 2.8

Mean TTR

Mean ln (TTR)	4.91
MTTR = $e^{4.91}$ (min)	136.17
136.17/60 (hr)	2.27

7. Calculate the upper confidence limit for time to repair in minutes and convert to hours, as shown in Table 2.9. The upper confidence limit is calculated as

$$\text{UCL}_{\text{LnTTR}} = \mu_{\text{LnTTR}} + t_{1-\alpha,v}\sigma_{\text{LnTTR}} \qquad (2.40)$$

where the Student's t sampling distribution is evaluated for $1 - \alpha$ and the degrees of freedom, $v = n - 1$. Assuming a 95% confidence limit,

TABLE 2.9

Upper Confidence Limit for TTR

$t_{1-\alpha,v}$	1.686	$T_{INV}(0.95, 39-1)$
$UCL_{\ln(TTR)}$	5.16	
$UCL_{TTR} = e^{5.16}$ (min)	173.65	
173.65/60 (hr)	2.89	

the upper confidence limit for the time-to-failure data is 2.89 hours, as shown in Table 2.9.

Weibull Approach

Fitting data to the Weibull repair math model was covered in the failure math modeling section in this chapter. The distinction for repair math modeling is calculating the upper confidence limit. This is readily performed in Excel using Goal Seek. The parameters of the Weibull distribution are found through median ranks regression and correlation analysis, as shown in Table 2.10.

TABLE 2.10

Weibull Repair Math Model Parameters: Summary Output

Regression Statistics					
r	0.998				
r^2	0.9967				
Adj r^2	0.9966				
SE[a]	0.070				
Obs	39				

ANOVA[b]	df	SS[c]	MS[d]	F	P
Regression	1	55.48	55.48	11174	0.000
Residual	37	0.18	0.00		
Total	38	55.66			

Coefficients		SE	t Stat	P	LCI[e] 95%	UCI[f] 95%
y_0	−5.1235	0.04	−114.76	0.000	−5.21	−5.03
β_3	1.4507	0.01	105.71	0.000	1.42	1.48
η_3	34.18	$e^{-(-5.1235/1.4507)}$				
γ	107					

[a] SE = standard error.
[b] ANOVA = analysis of variance.
[c] SS = sum of squares.
[d] MS = mean of squares.
[e] LCI = lower confidence interval.
[f] UCI = upper confidence interval.

TABLE 2.11

95% Upper Confidence Limit for TTR

Goal seek	
C	0.95
UCL_{TTR} (hr)	179.71 ($\pm$3.00)
$F(UCL)$ $(1 - exp(-(((UCL_{TTR} - \gamma)/\eta)^{\beta})))$	0.950

Goal seek finds the value for the upper confidence limit that meets the condition, $F(UCL) = 1 - \alpha$, or C. The procedure is as follows:

1. Enter the value for $1 - \alpha$, or C in a cell.
2. Enter an estimate for the upper confidence limit in a cell (the estimate for the upper confidence limit must be greater than the location parameter).
3. Enter the equation for the Weibull cumulative repair math model in a cell.

Goal seek presents a dialog box that sets the equation for the cumulative repair math model, $F(t)$, to the value for $1 - \alpha$ by changing the value estimated to be the upper confidence limit (Table 2.11).

System Logistics Downtime Math Model

The logistics downtime model is developed for the system, not a part (Figure 2.34). Logistics downtime begins with notification that a part failure has occurred causing a system downing event. The logistics events performed to transport the system to the maintenance facility, or transport the maintenance crew to the system, are designated as prerepair logistics downtime. Prerepair logistics downtime also includes all delays that prevent the maintenance action to begin. The delays include waiting for labor, access to the maintenance facility, specialty tools, spare parts, and administrative tasks. The logistics events performed to return a system to operations after the maintenance actions are complete are designated as postrepair logistics downtime. Postrepair logistics downtime includes system test runs, certification of repair actions, transport of the system to the operations location, and administrative tasks.

Triangular Distribution

Subjective statistical information fits the triangular distribution. The parameters of the triangular distribution are the minimum, mode, and maximum values for an event. Subjective probability and statistics make use of subject matter experts' (SME) knowledge to evaluate ranges of independent

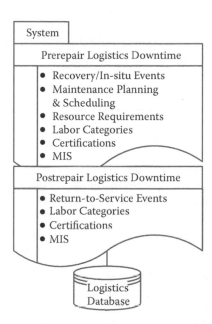

FIGURE 2.34
Qualitative logistics downtime analysis.

variables such as time. Subject matter experts identify the events required to perform prerepair and postrepair logistics and they conduct a Delphi survey that specifies the minimum, mode, and maximum time required to perform those events, as shown in Table 2.12. Logistics downtime events are characterized in minutes as with repair model data. The final characterization of the parameters of the triangular distribution will be in hours.

Event and Delphi Survey from a Subject Matter Expert

The findings from the Delphi survey for all subject matter experts are used to determine the parameters of the prerepair and postrepair logistics downtime math model as shown in Table 2.13 (time estimates, time totals, are initially in minutes and then converted to hours: 341 minutes for the sum of the minimal time for prerepair logistic downtime is converted to 5.68 hours). The estimates for the prerepair and postrepair logistics downtime minimum, mode, and maximum times are the mean of the time estimates provided by the subject matter expert.

The mean logistics downtime for prerepair and postrepair logistics events, Λ, is calculated as the arithmetic mean of the means for the minimum, mode, and maximum times. The triangular logistics downtime distribution and mean are expressed as follows:

The probability mass density function of the triangular distribution, $f(t)$, consists of four arguments: $f(t) = 0$ for all values of time, t, between 0 and T

TABLE 2.12

Logistics Downtime Events and Subject Matter Expert Delphi Survey Data

	Minutes		
Prerepair logistics events	T_{min}	T_{mode}	T_{max}
Failure notification	15	20	30
Fault detection/isolation	60	75	120
Doc—failure report	15	20	25
Spare part acquisition	45	60	90
Maintenance vehicle/specialty tool queue	45	75	180
Staffing for maintenance action	45	60	90
Sum (min)	225	310	535
λ (min)	356.67		
λ (hr)	5.94		
Postrepair logistics events			
Doc—corrective action	20	30	45
Doc—labor, materials	30	40	60
Doc—repair certification/calibration	20	30	35
Part/materials disposal	30	45	50
Return system to service	0	15	20
Sum (min)	100	160	210
λ (min)	156.67		
λ (hr)	2.61		

TABLE 2.13

Logistics Downtime Math Model Data

Prerepair Logistics Downtime				Postrepair Logistics Downtime			
	T_{min}	T_{mode}	T_{max}		T_{min}	T_{mode}	T_{max}
SME 1	69	90	194	SME 1	35	44	93
SME 2	73	109	176	SME 2	30	66	96
SME 3	66	122	176	SME 3	29	58	102
SME 4	61	111	183	Sum (min)	94	168	291
SME 5	72	95	184	Sum (hr)	1.57	2.80	4.85
Sum (min)	341	527	913	$\lambda_{postrep}$ (hr)			3.07
Sum (hr)	5.68	8.78	15.22				
λ_{prerep} (hr)			9.89				

minimum; $f(t)$ is fit to a line that begins at T minimum and ascends to the maximum frequency at T mode for all values of time between T minimum and T mode; $f(t)$ is fit to a line that begins at T mode and descends to T maximum for all values of time between T mode and T maximum; and $f(t) = 0$ for all values of time greater than T maximum.

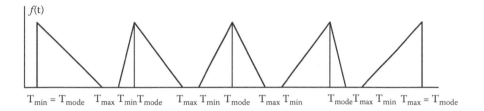

FIGURE 2.35
Range of shapes of the triangular distribution.

$$
f(t) = \begin{cases}
0 & \text{if} \quad 0 \le t < T_{min} \\
\dfrac{2(t - T_{min})}{(T_{max} - T_{Min})(T_{mode} - T_{min})} & \text{if} \quad T_{min} \le t < T_{mode} \\
\dfrac{2(T_{max} - t)}{(T_{max} - T_{Min})(T_{max} - T_{mode})} & \text{if} \quad T_{mode} \le t \le T_{max} \\
0 & \text{if} \quad t > T_{max}
\end{cases} \tag{2.41}
$$

The mean of the triangular distribution is the sum of the three values for time divided by three as shown in the following equation.

$$
\mu_\Lambda = \frac{T_{min} + T_{mode} + T_{max}}{3} \tag{2.42}
$$

The shape of the triangular distribution mass density function can range from a right triangle skewed to the right, a positively skewed distribution; asymmetrical distribution about the mean, a negatively skewed distribution; and a right triangle skewed to the left, as shown in Figure 2.35.

Summary of Reliability Math Models

Reliability math models for time to failure, prerepair logistics downtime, time to repair, and postrepair logistics downtime describe the operational and maintenance cycle of a part over the useful life of a system. The math models are presented in a timeline for notional shapes in Figure 2.36. The time-to-failure math model describes the expectation of the incidence of part failure. The prerepair logistics downtime model describes the expectation for the duration of prerepair logistics events immediately following part failure. The time-to-repair math model describes the expectation of the duration of maintenance events. The postrepair logistics downtime model describes

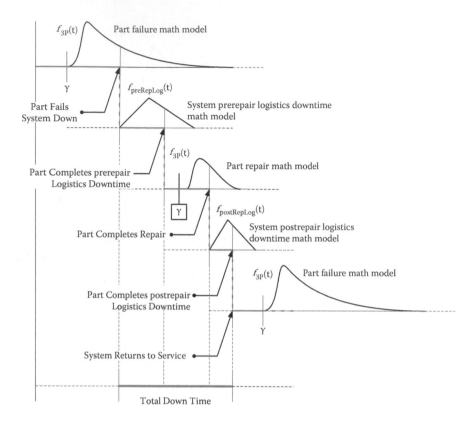

FIGURE 2.36
Part failure maintenance cycle reliability math models.

the expectation for the duration of postrepair logistics events immediately following the completion of maintenance actions.

Availability

Availability is the probability that a part, an assembly, up to the system design configuration will be able to perform its mission when required. The basic equation for availability is uptime divided by the sum of uptime and downtime. Availability for higher-level design configurations is calculated from part availability in the same way that reliability of higher-level configurations is calculated from part reliability. Design engineers are called upon to calculate inherent and operational availability for parts. The distinction between inherent and operational availability is based on the definition of uptime and downtime.

Inherent Availability, A_i

Inherent availability is a design metric that predicts the ideal availability for a part. The calculation for inherent availability for a part is expressed in the following equation:

$$A_i = \frac{\text{MTBF}}{\text{MTBF} + \text{MTTR}} \tag{2.43}$$

Uptime for the inherent availability is defined as the part mean time between failure (MTBF). Downtime for the inherent availability is defined as the part mean time to repair (MTTR). Recall that the repair experiments performed by design engineers do not exactly replicate the conditions for repair of the part by the user. If we assume that the design MTTR is close to the realized MTTR by the user, then the inherent availability calculated by the design engineer can be viewed as the ideal availability of the part that will not be exceeded in the field by the user.

The inherent availability of an assembly is determined by the design configuration of the assembly. The inherent availability of an assembly will be equal to the product of the inherent availability of the parts for a serial design configuration. The inherent availability assembly for redundant design configurations can be calculated from the equations presented for the survival function and mission reliability functions in the preceding section.

Operational Availability, A_O

Operational availability is a predictive metric calculated by the design engineer. The calculation for operational availability for a part is expressed in the following equation:

$$A_O = \frac{\text{MTBF}}{\text{MTBF} + \text{MTTR} + \text{MLDT}} \tag{2.44}$$

Uptime for operational availability is defined as the part MTBF, as for the inherent availability. Downtime for the operational availability is defined as the sum of the MTTR, and the mean logistics downtime. The mean logistics downtime is equal to the sum of the mean prerepair logistics downtime and the mean postrepair logistics downtime. Logistical downtime is often presented as a single value designated administrative and logistics downtime in many texts and papers. The estimate for logistical downtime is typically a constant rather than the result of a qualitative and quantitative analysis of a distribution. Design engineers often lack the budget and schedule to conduct a Delphi experiment to fit the logistics downtime math models. More important, design engineers have no way of knowing what the prerepair and postrepair logistics downtime events are for each and every user. The deterministic estimate for logistical downtime is often the best information available.

Application of Reliability Math Models

The development of the reliability math models for parts provides the information that is used to develop reliability simulations that use the distributions in a Monte Carlo random variable generator to characterize the operational and maintenance cycles over the useful life of the asset for parts. The reliability simulations characterize the behavior of the part operational and maintenance cycles to describe the effects on the reliability, maintainability, and availability of the assembly, and all higher design configurations to include the system. The logic of reliability simulation is provided in Figure 2.37. Reliability simulation is addressed in Chapter 9, Reliability Simulation and Analysis.

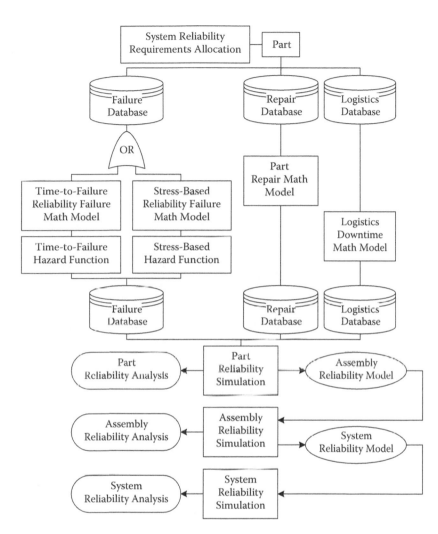

FIGURE 2.37
Quantitative reliability analyses inputs to reliability simulation.

3

Reliability Analysis for System Sustainment

Reliability analysis for system sustainment, in a very limited scope, has been an accepted best maintenance practice since 1962, with the advent of reliability-centered maintenance for the Boeing 747. However, a review of the literature yields very little information on how reliability analysis is implemented by system end users.

- A survey of reliability books finds that the emphasis is on statistical analysis that leads to the development of part failure math models, part repair math models, and reliability math models for design configurations.
- A survey of journal articles finds that the emphasis is on scholarly analysis that defies application by the reliability practitioner in the field supporting the end user and its system.
- A survey of professional development tutorials and professional society technical sessions finds that the emphasis is on anecdotal implementation of reliability analysis primarily for part design and rarely for part sustainability.

The only exceptions to the surveys of the literature are topics on reliability-centered maintenance. Chapter 7 covers reliability-centered maintenance. Reliability-centered maintenance is performed by the system end user, not the designer. The reliability analysis approach for reliability-centered maintenance differs significantly from the reliability analysis approach presented in the previous chapter for system design. The key reasons are as follows:

- The end user has no influence on the design analysis for the system: (1) the system design is locked in once the end user acquires it, and (2) the system design configuration is known.
- The end user is well aware of the critical parts that have life-cycle economic and safety impacts on the system; therefore a criticality analysis is not required.
- The end user has historical data for all critical parts that documents: (1) time to failure, (2) prerepair and postrepair logistics downtime, and (3) time to repair; therefore a qualitative failure, repair, and logistical downtime, analysis is not required.

Reliability analysis for system sustainment begins with a known critical part and applies the quantitative failure, repair, and logistics downtime analyses that will be performed by the end user to (1) understand the current baseline reliability functions of the critical parts and to (2) implement sustainment practices that will improve the reliability functions of the critical parts.

Baseline Reliability Analysis

Reliability analysis for a critical part requires information:

- Time to failure for the end user's population of a specific critical part
- Time to repair for the end user's population of a specific critical part
- Duration of prerepair and postrepair logistics downtime for the end user's population of a specific critical part

Development of the baseline reliability analysis requires data. The sources of data are historical maintenance and operating records of the organization. Every organization captures data in a management information system in their own way. For accounting, the minimum amount of data for recording is labor, materials, and overhead expenses, and for maintenance planning and scheduling. The question that must be answered is whether the content of the management information system is capable of providing the information required for a reliability analysis. The management information system must be evaluated to determine whether it is capable of providing the time to failure, time to repair, and duration of logistical downtime for critical parts. An approach to performing this evaluation begins with determination of the information that is necessary to perform the baseline reliability analysis. Let us, for lack of a better term, refer to the reliability information acquisition system as the "failure reporting, analysis, and corrective actions system" (FRACAS).

Development of the baseline reliability analysis also includes documenting the existing maintenance practices that are performed by the organization and the existing logistics practices. Both of these tasks are performed concurrently with development of the failure reporting, analysis, and corrective action system.

The reliability information that currently exists is used to fit the critical part's failure and repair math models, and the system logistics downtime math model. The critical part's failure, repair, and logistics downtime math models are entered into a reliability simulation program. Life-cycle simulations for the critical part enable analysis of the baseline reliability functions. All of the information that is gathered from the management information

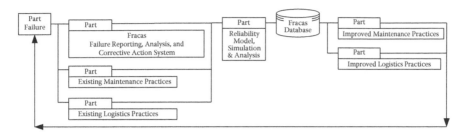

FIGURE 3.1
FRACAS flow chart.

system, reliability math models, and life-cycle simulation reports is entered into the failure reporting and corrective actions system database.

Information from the life-cycle simulation reports enables maintenance personnel to investigate and implement improvement alternatives for maintenance and logistics practices. The revised maintenance and logistics practices become the first revision to the baseline maintenance and logistics practices.

The reliability analysis is repeated as improvement alternatives are implemented and critical parts fail. Subsequent revisions to the baseline maintenance and logistics practices are documented as policy as new improvement alternatives are implemented. Therefore the reliability analysis is a closed loop system for continuous improvement, as illustrated in Figure 3.1.

Failure Report, Analysis, Corrective Action System—FRACAS

Part failure reporting, root-cause failure analysis, and corrective action are essential maintenance practices for reliability, safety, and logistics analysis. This process has been formalized by organizations in forms that fit their maintenance management model. The U.S. Department of Defense formalized its process for failure reporting, analysis, and corrective action system (FRACAS). Their FRACAS is a closed loop management information system that defines a process for reporting, classifying, and analyzing failures and planning corrective actions in response to those failures. Defense FRACAS standards and handbooks provide extensive information on form format and failure evaluation procedures that apply to the military system acquisition in which the system sustainer (the Department of Defense) and the vendor (prime contractors) participate in a cradle-to-grave collaboration for the design, development, test and evaluation, manufacture, transport, storage, operation, and maintenance of systems.

The cradle-to-grave collaboration does not exist for industrial and consumer products. Typical relationships between an organization that designs and develops a system with the end user are (1) marketing and sales functions where the design organization seeks to know the "voice of the customer" (a collaboration), and (2) the warranty failures where the system fails to provide the end user with the promised utility (a confrontation). More often, there is no relationship between the designer and the end user due to the complexity of the distribution system. End users have a direct relationship with the retail organization that sold the system and may provide maintainability support. Retailers have several distribution organizations between them and the system designer that impede or prevent any direct relationship with the system designer. End users have sole responsibility to sustain the asset with minimal or no supporting reliability information from the designer.

Information gathered by FRACAS must include:

- An accurate measurement of the time to failure for the critical part

  For example, consider the air brake canister for a mine haul truck. Each haul truck has two air brake canisters. The mine operates twenty haul trucks. Forty air brake canisters are in use on any given day. Assume that fifty air brake canisters have failed over the past five years. An accurate measurement of the time to failure for the air brake canister would include the operating hours of each air brake canister from the time of its installation to the time that it failed.

- An accurate measurement of the time to repair for the critical part

  An accurate measurement of the time to repair for the air brake canister will include the time that the repair began and the time that the repair was concluded for each air brake canister. Time-to-repair data must be qualified by the maintenance environment in which the repair was performed. Consider two scenarios:

  1. The air brake canister was removed and replaced in a maintenance bay in a controlled and sheltered environment after the haul truck had been steam cleaned and allowed to cool down before the repair was performed.

  2. A maintenance crew travels to the haul truck located in the mine at 2 o'clock in the morning in February during a sleet storm, or the haul truck is located in the mine at 2 o'clock in the afternoon in August and is exposed to dust and debris.

  Assume that the frequencies of maintenance actions in the two scenarios are equal. The time-to-repair math model for the air brake canister calculated for the fifty occurrences will be flawed. This scenario represents two independent and mutually exclusive populations that have unique time-to-repair math models. [NOTE: a benefit of reliability-based sustainability is the minimization of the second scenario by enabling an organization to implement a preventive maintenance practice.]

- An accurate measurement of the logistics downtime for the system

The logistics downtime for a system is typically the same duration for any critical part failure. Therefore a part does not have a logistics downtime. Logistics downtime occurs between notification of the critical part failure and the commencement of repair maintenance events, and between conclusion of repair maintenance events and the return of the system to operations. The former logistics downtime is referred to as prerepair logistics downtime and the latter is referred to as postrepair logistics downtime. The logistics events and duration for the two represent independent and mutually exclusive populations. An accurate measurement of each must include identification of the events that define the duration.

Prerepair Logistics Downtime

- Time required to move the system to the maintenance facility, or time required to transport the maintenance personnel to the system
- Time required to assign and staff maintenance personnel to perform fault detection and fault isolation
- Waiting time for access to a maintenance bay
- Waiting time for access to a specialty tool, i.e., an overhead crane, diagnostic equipment, etc.
- Waiting time for spare parts and associated hardware
- Waiting time for system cool down
- System preparation, i.e., steam cleaning
- Administrative procedures, i.e., failure reports, maintenance planning reports, etc.

Postrepair Logistics Downtime

- Trial operation to assure the system repair is valid
- Servicing to replenish expended fluids, recharge compressor tank, etc.
- The transportation of system to operations
- Administrative procedures, i.e., failure reports, labor, materials, overhead accounting, etc.

Quantitative Reliability Analysis

Quantitative reliability analysis fits the data for time to failure, time to repair, and the durations of prerepair and postrepair logistics downtime to math models. The source of the information for development of the math models is the failure reporting, analysis, and corrective action system (FRACAS) database, as depicted in Figure 3.2.

Failure Math Model

Field sustainment reliability analyses fit time-to-failure math models. Units of time can be chronological, engine hours, cycles, cold starts, etc. A key distinction between design reliability analysis failure math models and sustainment reliability analysis math models is that all sustainment math models

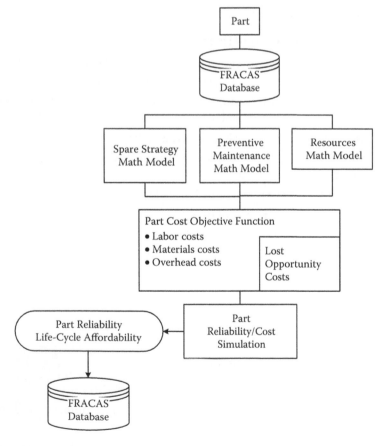

FIGURE 3.2
Quantitative reliability analysis.

are time censored. A time-censored set of time-to-failure data includes the time in service for all of the surviving critical parts at the time that the analysis is performed. Consider the example of the air brake canister for the mine haul trucks. The reliability analysis to fit the time-to-failure math model is performed for all data on a specific date. The FRACAS database provides the time to failure for every air brake canister that has been replaced for the fleet of haul trucks. On that date, forty air brake canisters are in use and have accumulated failure-free operating time since the date of their installation. The total time on test for the air brake canisters is equal to the sum of the time to failure for all of the replaced air brake canisters, plus the accumulated failure-free operating time for all of the surviving air brake canisters.

Reliability Functions

Quantitative part failure analysis fits math models that describe the reliability functions. The procedure to express the reliability functions includes the following:

1. Descriptive statistics and frequency distribution of sample data

   The time-to-failure data are recorded in a spreadsheet. The descriptive statistics of the data are calculated to include the sample mean and standard deviation, the sample kurtosis and skewness, and the sample size. The data are fit to a frequency distribution that plots the sample histogram. The sample histogram describes the approximate shape of the distribution of the data: exponential, skewed, or symmetrical about the mean. The accumulated failure-free time for the surviving critical parts is not included.

2. Failure probability distribution, pdf, $f(t)$

   The data are fit to a probability mass density distribution, the pdf, expressed as $f(t)$, for time-to-failure data. The failure probability distribution describes the behavior of the failure mode over time.

3. Cumulative failure probability distribution, cdf, $F(t)$

   The cumulative probability mass density distribution, the cdf, expressed as $F(t)$, for time-to-failure data is calculated as the integral of $f(t)$ from 0 to t. The cumulative failure probability distribution describes the cumulative proportion of failure over time. The cumulative failure probability distribution is equal to 0 at $t = 0$. The cumulative failure probability distribution approaches 1 as t approaches infinity.

4. Survival function, $S(t)$

   The survival function, expressed as $S(t)$, for time-to-failure data, is calculated as the complement of the cumulative probability distribution, $S(t) = 1 - F(t)$. The survival function is equal to 1 at $t = 0$. The survival function approaches 0 as t approaches infinity. The indefinite integral of the part survival function is equal to the mean time between failure for a part; indefinite integral of the system survival function is equal to the mean time between downing events for a system.

5. Mission reliability function, $R(\tau|t)$ given $\tau \equiv$ mission duration

   The mission reliability function for time-to-failure data, expressed as $R(\tau|t)$, is the conditional probability that a part will perform failure free for the next mission duration given that it has survived to the current time.

6. Hazard Function, $h(t)$

   The hazard function is the instantaneous failure rate of the part over time. The hazard function is the ratio of the failure probability distribution and the survival function, and is expressed as $h(t) = f(t)/S(t)$.

Fitting Reliability Data to Reliability Math Models

Commercially available reliability simulation software programs require that the engineer identify and fit failure, repair, and logistical downtime math models from empirical data. Two reliability failure math models are presented: the exponential and the Weibull distributions. Two reliability repair math models are presented: the log-normal and the Weibull distributions. One repair logistics downtime math model is presented: the triangular distribution. The methods used to fit the math models are presented in detail in Appendix A.

Exponential Reliability Math Models

Approach for Complete Data

The exponential reliability math model is fit by the arithmetic mean of the time-to-failure data, expressed in chronological time, hours, cold starts, etc. In Table 3.1, time is measured as cold starts not unlike the starter motor of an automobile. The mean time between failure is twenty missions. The failure rate is the inverse of the mean time between failure, 0.05 failures per mission. The mission duration is one start per mission.

Table 3.2 describes how the exponential failure math model behaves. Assume that 100 systems, N_0, are available at the beginning of its useful life. At the conclusion of the first mission, five parts will fail ($r = \lambda N_0$), $f(t)$. The cumulative failure distribution, $F(t)$, is equal to the five failed parts, $F(t) = \Sigma f(t)$. The survival function, $S(t)$, is equal to the ninety-five parts that have not failed

TABLE 3.1

Exponential Reliability
Failure Math Model

θ (missions)	20
λ (fpm)	0.05
τ (missions)	1

TABLE 3.2

Exponential Failure Math Model Data

m	N	r	F(m)	S(m)	R(m)
0	100	0	0	100	0.95
1	95.00	5.00	5.00	95	0.95
2	90.25	4.75	9.75	90.25	0.95
3	85.74	4.51	14.26	85.74	0.95
4	81.45	4.29	18.55	81.45	0.95
5	77.38	4.07	22.62	77.38	0.95
6	73.51	3.87	26.49	73.51	0.95
7	69.83	3.68	30.17	69.83	0.95
8	66.34	3.49	33.66	66.34	0.95
9	63.02	3.32	36.98	63.02	0.95
10	59.87	3.15	40.13	59.87	0.95
11	56.88	2.99	43.12	56.88	0.95
12	54.04	2.84	45.96	54.04	0.95
13	51.33	2.70	48.67	51.33	0.95
14	48.77	2.57	51.23	48.77	0.95
15	46.33	2.44	53.67	46.33	0.95
16	44.01	2.32	55.99	44.01	0.95
17	41.81	2.20	58.19	41.81	0.95
18	39.72	2.09	60.28	39.72	0.95
19	37.74	1.99	62.26	37.74	0.95
20	35.85	1.89	64.15	35.85	0.95
21	34.06	1.79	65.94	34.06	0.95
22	32.35	1.70	67.65	32.35	0.95
23	30.74	1.62	69.26	30.74	0.95
24	29.20	1.54	70.80	29.20	0.95
25	27.74	1.46	72.26	27.74	

$(1 − F(t))$. The mission reliability is equal to the survival function at the conclusion of the first mission divided by the survival function at the beginning of the useful life of the system, $S(t_1)/S(t_0)$. At the conclusion of the second mission, 4.75 parts will fail ($r = \lambda N_1$). The cumulative failure distribution, $F(t)$, is equal to the 9.75 failed parts, $F(t) = \Sigma f(t)$. The survival function, $S(t)$, is equal to the 90.25 parts that have not failed $(1 - F(t))$. The mission reliability is equal to the survival function at the conclusion of the first mission divided by the survival function at the beginning of the useful life of the system, $S(t_2)/S(t_1)$. Observe that the mission reliability function remains constant for all missions. Also note in Table 3.2 that sixty-four parts have failed by the time the system reaches the mean time between failure of twenty missions.

The frequency distribution for the sample data is presented in Figure 3.3. It represents the expected shape of an exponential distribution.

The cumulative failure distribution and survival function is presented in Figure 3.4. Observe that they intersect at the median time–time to fail. Fifty percent of the parts will have failed by the conclusion of the 14th mission.

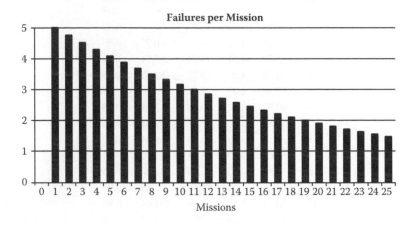

FIGURE 3.3
Exponential sample frequency distribution histogram.

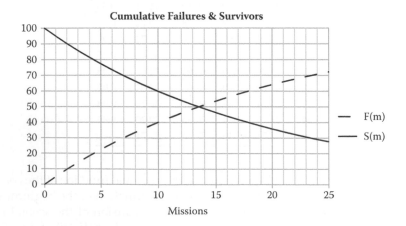

FIGURE 3.4
Exponential cumulative failure distribution and survival function.

Approach for Time-Censored Data

Time-censored failure data cannot be explained by the graphical approach for sample failures in time expressed for complete failure data. Instead, the objective is to take the data that exist on the date of the analysis and calculate the parameters of the exponential failure math model and its lower confidence limit for the mean time between failure.

Recall that the mean time between failure for complete failure data is equal to the arithmetic mean of the times to failure. The mean time between failure for time-censored failure data is equal to the total time on test divided by the total number of failed parts.

The total time on test, T, is equal to the summation of all of the times to failure plus the accumulated time in service for the surviving parts. This is expressed in the following equation:

$$T = \frac{\sum_{i=1}^{r} T_i + \sum_{j=1}^{n} T_j}{r} \tag{3.1}$$

where
   T_i is defined as the time to failure for the ith part
   r is defined as the total number of failed parts
   T_j is defined as the failure-free time for the jth part
   n is defined as the total number of surviving parts at the time the calculation is performed

The mean time between failure, θ, is equal to the total time, T, divided by the total number of failed parts, r, as expressed in the following equation:

$$\theta = \frac{T}{r} \tag{3.2}$$

The lower confidence limit for the mean time between failure, θ_{LCL}, for time-censored failure data is calculated using the total time, T, and the chi-squared distribution evaluated for the level of significance, α, and $2r + 2$ degrees of freedom, as expressed in the following equation:

$$\theta_{LCL} = \frac{2T}{\chi^2_{\alpha,2r+2}} \tag{3.3}$$

Exponential Part Failure Math Model

The exponential part failure math model for the data is fit by the mean time between failure, as shown in the following equation. The exponential part failure math model is typically expressed in the alternate form using the failure rate rather than the mean time between failure.

$$f(t) = \frac{1}{\theta} e^{-\left(\frac{t}{\theta}\right)} \tag{3.4}$$

$$\lambda = \frac{1}{\theta} \tag{3.5}$$

$$f(t) = \lambda e^{-\lambda t} \tag{3.6}$$

Exponential Cumulative Part Failure Math Model

The exponential cumulative part failure math model is found by the integral of the part failure math model, as shown in the following equation.

$$F(t) = 1 - e^{-\lambda t} \tag{3.7}$$

Exponential Part Survival Function

The exponential survival function is the complement of the cumulative part failure math model, as shown in the following equation.

$$S(t) = e^{-\lambda t} \tag{3.8}$$

Exponential Part Mission Reliability Function

The exponential part mission reliability function is the conditional probability that the part will survive the next mission given that it has survived to time, t. It is shown that the exponential part mission reliability function is constant for specified values for the mission duration since the failure rate and mission duration are constant.

$$R(\tau \mid t) = \frac{S(t + \tau)}{S(t)} = \frac{e^{-\lambda(t+\tau)}}{e^{-\lambda t}} = \frac{e^{-\lambda t} e^{-\lambda \tau}}{e^{-\lambda t}} = e^{-\lambda t} \tag{3.9}$$

$$R(\tau \mid t) = S(\tau) \tag{3.10}$$

In many texts and papers, the survival function is expressed as $R(t)$ and called the life function. This is because the exponential mission reliability function is equal to the survival function evaluated for the mission duration.

A part will have a constant reliability until the mission in which it fails, according to the exponential mission reliability function.

Exponential Part Hazard Function

The exponential part hazard function is its constant failure rate. The failure rate applies equally to parts on its first mission, the tenth, the hundredth, ad infinitum.

$$h(t) = \frac{f(t)}{1 - F(t)} = \frac{f(t)}{S(t)} = \frac{\lambda e^{-\lambda t}}{e^{-\lambda t}} = \lambda \tag{3.11}$$

Weibull Reliability Math Models

Approach for Complete Data

The following example for time-to-failure data presents a scenario that cannot be fit by the exponential failure math model but can be fit to the Weibull failure math model (Table 3.3). The time units are cold starts per mission, just like the exponential example. Assume that 100 systems, N_0, are available at the beginning of their useful lives. Notice that the incidence of part failures begins at the 17th mission. Until the 17th mission the part failure math model, $f(t)$, is 0, the cumulative failure math model is 0, the survival function is 1.0, and the mission reliability is 1.0.

At the conclusion of the 17th mission, two parts will fail, $f(t)$. The cumulative failure distribution, $F(t)$, is equal to the two failed parts, $F(t) = \Sigma f(t)$. The survival function, $S(t)$, is equal to the ninety-eight parts that have not failed $(1 - F(t))$. The mission reliability is equal to the survival function at the conclusion of the 17th mission divided by the survival function at the 16th mission, $S(t_{17})/S(t_{16})$. The hazard function, $h(m)$, is equal to the number of failures that occur on the 17th mission divided by the number of survivors at the completion of the 16th mission, (r_{17}/S_{16}). At the conclusion of the 18th mission, nine parts fail. The cumulative failure distribution, $F(t)$, is equal to the eleven failed parts, $F(t) = \Sigma f(t)$. The survival function, $S(t)$, is equal to the eighty-nine parts that have not failed $(1 - F(t))$. The mission reliability is equal to the survival function at the conclusion of the 18th mission divided by the survival function 17th mission, $S(t_{18})/S(t_{17})$. Observe that the mission reliability function declines following the incidence of the first failure. Also note that the hazard function increases following the incidence of the first failure.

The frequency distribution for the sample data is presented in Figure 3.5. It represents a negatively skewed frequency distribution. The lack of part failures before the 17th mission suggests that a location parameter exists.

TABLE 3.3

Weibull Reliability Failure Math Model Data

m	N	r	F(m)	S(m)	R(m)	h(m)
0	100	0	0	100		0
1	100	0	0	100	1	0
2	100	0	0	100	1	0
3	100	0	0	100	1	0
4	100	0	0	100	1	0
5	100	0	0	100	1	0
6	100	0	0	100	1	0
7	100	0	0	100	1	0
8	100	0	0	100	1	0
9	100	0	0	100	1	0
10	100	0	0	100	1	0
11	100	0	0	100	1	0
12	100	0	0	100	1	0
13	100	0	0	100	1	0
14	100	0	0	100	1	0
15	100	0	0	100	1	0
16	100	0	0	100	1	0
17	98	2	2	98	0.980	0.020
18	89	9	11	89	0.908	0.092
19	71	18	29	71	0.798	0.202
20	38	33	62	38	0.535	0.465
21	8	30	92	8	0.211	0.789
22	0	8	100	0	0	1
23	0	0	100	0		
24	0	0	100	0		
25	0	0	100	0		

FIGURE 3.5

Weibull sample frequency distribution histogram.

FIGURE 3.6
Weibull sample cumulative failure distribution and survival function.

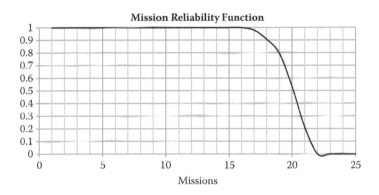

FIGURE 3.7
Weibull mission reliability function.

The cumulative failure distribution and survival functions are presented in Figure 3.6. Observe that they intersect at the median time–time to fail. Fifty percent of the parts will have failed by the conclusion of the 20th mission.

The mission reliability function is presented in Figure 3.7. Observe that the mission reliability is 1 until the 17th mission and then decreases over time.

The hazard function is presented in Figure 3.8. Observe that the hazard function is 0 until the 17th mission and then increases over time.

Approach for Time-Censored Data

Just as for the exponential distribution, time-censored failure data cannot be explained by the graphic approach for sample failures in time expressed for complete failure data. Median ranks regression uses data that exist on the date of the analysis to fit the parameters of the Weibull failure math as presented in Appendix A.

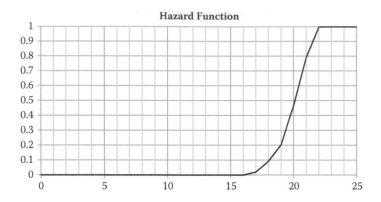

FIGURE 3.8
Weibull hazard function from sample data.

Weibull Part Failure Math Model

The Weibull part failure math model for the data is fit from the failure data by the median ranks regression and coefficient correlation analysis. The Weibull failure math model is expressed in the following equation:

$$f(t) = \begin{vmatrix} 0 & \text{for} & t < \gamma \\ \left(\dfrac{\beta}{\eta^{\beta}}\right)(t-\gamma)^{\beta-1} e^{-\left(\frac{t-\gamma}{\eta}\right)^{\beta}} & \text{for} & t \geq \lambda \end{vmatrix} \tag{3.12}$$

The following examples show the effects of the shape parameter, β, and the location parameter, γ, on the failure math model (Figure 3.9). Both Weibull failure math models, $f_1(t)$ and $f_2(t)$, have equal values for the scale parameter, η. The first shape parameter, β_1, is 2.25 with a location parameter, γ, equal to 0. The Weibull failure math model for β_1 and $\gamma = 0$, is the positively skewed solid black line, $f_1(t)$. The Weibull failure math model for β_2 and $\gamma = 25$, is the negatively skewed dashed black line, $f_2(t)$.

Weibull Cumulative Part Failure Math Model

The Weibull cumulative part failure math model is found by the integral of the part failure math model, as shown in the following equation.

$$F(t) = \begin{vmatrix} 0 & \text{for} & t < \gamma \\ 1 - e^{-\left(\frac{t-\gamma}{\eta}\right)^{\beta}} & \text{for} & t \geq \lambda \end{vmatrix} \tag{3.13}$$

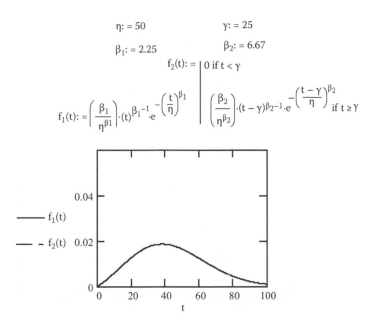

$\eta: = 50$

$\gamma: = 25$

$\beta_1: = 2.25$

$\beta_2: = 6.67$

$$f_2(t): = \begin{vmatrix} 0 \text{ if } t < \gamma \\ \left(\dfrac{\beta_2}{\eta^{\beta_2}} \right) \cdot (t - \gamma)^{\beta_2 - 1} \cdot e^{-\left(\dfrac{t - \gamma}{\eta} \right)^{\beta_2}} \text{ if } t \geq \gamma \end{vmatrix}$$

$$f_1(t): = \left(\dfrac{\beta_1}{\eta^{\beta_1}} \right) \cdot (t)^{\beta_1 - 1} \cdot e^{-\left(\dfrac{t}{\eta} \right)^{\beta_1}}$$

$f_1(t)$ ———

$f_2(t)$ — —

FIGURE 3.9
Weibull failure math model—shape parameter.

The following examples show the effects of the shape parameter, β, and the location parameter, γ, on the cumulative failure math model (Figure 3.10). The Weibull cumulative failure math model for β_1 and $\gamma = 0$, is the solid black line, $F_1(t)$, that originates at the origin and approaches 1. The Weibull failure math model for β_2 and $\gamma = 25$, is the dashed black line, $F_2(t)$, that originates at the location parameter and approaches 1.

Weibull Part Survival Function

The Weibull survival function is the complement of the cumulative failure math model, as shown in the following equation.

$$S(t) = \begin{vmatrix} 1 & \text{for} & t < \gamma \\ e^{-\left(\frac{t - \gamma}{\eta} \right)^{\beta}} & \text{for} & t \geq \lambda \end{vmatrix} \tag{3.14}$$

The following examples show the effects of the shape parameter, β, and the location parameter, γ, on the survival function (Figure 3.11). The Weibull survival function for β_1 and $\gamma = 0$, is the solid black line, $S_1(t)$, which originates at 1 and approaches 0. The Weibull failure math model for β_2 and $\gamma = 25$, is the dashed black line, $S_2(t)$, that originates at the location parameter and approaches 0.

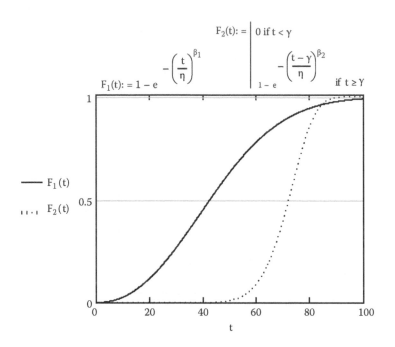

$$F_2(t) := \left| \begin{array}{l} 0 \text{ if } t < \gamma \\ \\ 1 - e^{-\left(\frac{t-\gamma}{\eta}\right)^{\beta_2}} \quad \text{if } t \geq \gamma \end{array} \right.$$

$$F_1(t) := 1 - e^{-\left(\frac{t}{\eta}\right)^{\beta_1}}$$

FIGURE 3.10
Weibull failure math model—location parameter.

Weibull Part Mission Reliability Function

The Weibull mission reliability function is the conditional probability for the survival function evaluated for the mission duration given that the part has survived to time, t, as shown in the following equation.

$$R(t) = \left| \begin{array}{ll} 1 & \text{for} \quad t < \gamma \\ \\ \dfrac{e^{-\left(\frac{t-\gamma+\tau}{\eta}\right)^{\beta}}}{e^{-\left(\frac{t-\gamma}{\eta}\right)^{\beta}}} & \text{for} \quad t \geq \lambda \end{array} \right. \tag{3.15}$$

The following examples show the effects of the shape parameter, β, and the location parameter, γ, on the mission reliability function (Figure 3.12). The Weibull mission reliability function for β_1 and $\gamma = 0$, is the solid black line, $R_1(\tau|t)$, that originates at 1 and approaches 0. The Weibull mission reliability for β_2 and $\gamma = 25$, is the dashed black line, $R_2(\tau|t)$, which originates at the location parameter and approaches 0.

$$S_1(t) := e^{-\left(\dfrac{t}{\eta}\right)^{\beta_1}}$$

$$S_2(t) := \begin{vmatrix} 1 \text{ if } t < \gamma \\[12pt] e^{-\left(\dfrac{t-\gamma}{\eta}\right)^{\beta_2}} \text{ if } t \geq \gamma \end{vmatrix}$$

$$\theta_1 := \int_0^\infty S_1(t)\, dt = 44.287 \qquad \theta_2 := \int_0^\infty S_2(t)\, dt = 71.653$$

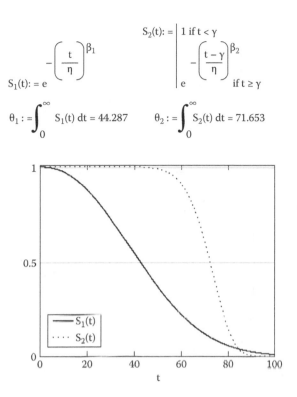

FIGURE 3.11
Weibull survival function and MTBF.

Weibull Part Hazard Function

The Weibull hazard function is the ratio of the Weibull failure math model, $f(t)$, and the survival function, $S(t)$, as shown in the following equation.

$$h(t) = \begin{vmatrix} 0 & \text{for} & t < \gamma \\[12pt] \left(\dfrac{\beta}{\eta^\beta}\right)(t-\gamma)^{\beta-1} & \text{for} & t \geq \lambda \end{vmatrix} \tag{3.16}$$

The following examples show the effects of the shape parameter, β, and the location parameter, γ, on the hazard function (Figure 3.13). The Weibull hazard function for β_1 and $\gamma = 0$, is the solid black line, $h_1(t)$, which originates at 0 and approaches 1. The Weibull hazard for β_2 and $\gamma = 25$ is the dashed black line, $h_2(t)$, which originates at the location parameter and approaches 1.

The hazard function is best plotted on logarithmic-linear (natural log, Naperian) scales due to the very small magnitude for $h(t)$.

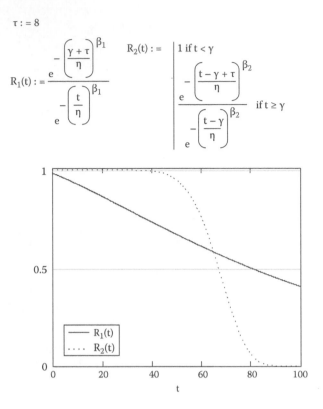

$$\tau := 8$$

$$R_1(t) := \frac{e^{-\left(\frac{\gamma + \tau}{\eta}\right)^{\beta_1}}}{e^{-\left(\frac{t}{\eta}\right)^{\beta_1}}}$$

$$R_2(t) := \begin{vmatrix} 1 \text{ if } t < \gamma \\[2ex] \dfrac{e^{-\left(\frac{t - \gamma + \tau}{\eta}\right)^{\beta_2}}}{e^{-\left(\frac{t - \gamma}{\eta}\right)^{\beta_2}}} & \text{if } t \geq \gamma \end{vmatrix}$$

FIGURE 3.12
Weibull mission reliability function.

Quantitative Maintainability Analysis

Mean Time to Repair

The basic building block for quantitative maintainability analysis is the part repair math model. The part repair math model is fit from empirical time-to-repair data that are acquired from historical records and repair experiments. Time to repair is the sum of the measures of the durations to perform maintenance repair events. The data for time to repair are typically positively skewed. The log-normal distribution has been the long standing best practice for fitting time-to-repair data to a repair math model. However, just as the Weibull distribution can fit skewed time-to-failure data, so too can it fit time-to-repair data.

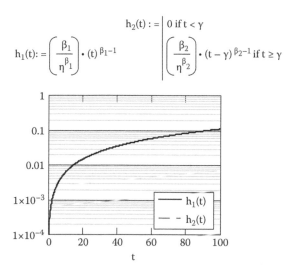

$$h_2(t) := \begin{vmatrix} 0 \text{ if } t < \gamma \\ \left(\dfrac{\beta_2}{\eta^{\beta_2}}\right) \cdot (t-\gamma)^{\beta_2-1} \text{ if } t \geq \gamma \end{vmatrix}$$

$$h_1(t) := \left(\dfrac{\beta_1}{\eta^{\beta_1}}\right) \cdot (t)^{\beta_1-1}$$

FIGURE 3.13
Weibull hazard function.

Maintainability Metrics

The basic design maintainability metric is the mean time to repair (MTTR). The MTTR is found for the repair math model. The upper confidence limit for time to repair is calculated to understand the behavior of time to repair.

Log-Normal Approach

Time-to-repair data are measured in hours. But repair events often occur in minutes. Fitting the repair math model in minutes often provides better accuracy for mean time to repair, and the upper confidence limit that can be converted to hours.

The procedure for fitting TTR data to the log-normal distribution is as follows:

1. Record the TTR data in Excel, as shown in Table 3.4.
2. Calculate the descriptive statistics, to include the sample mean and standard deviation, the kurtosis and skewness, and the sample size, as shown in Table 3.5.
3. Plot the sample TTR frequency distribution and histogram, as shown in Figure 3.14.

The histogram illustrates the positive skew of the TTR data.

TABLE 3.4

Time-to-Repair Data in Minutes

121	132	147
127	134	156
139	116	149
147	140	114
124	161	124
156	139	137
128	109	121
119	126	145
116	113	111
171	126	130
132	134	131
183	163	191
117	165	172

TABLE 3.5

Time-to-Repair Descriptive Statistics

X_{bar} (min)	137.59
s (min)	20.63
Kurtosis	0.054
Skewness	0.833
n	39
X_{min} (min)	109
X_{max} (min)	191

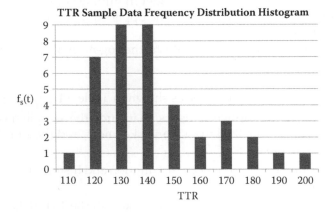

FIGURE 3.14
TTR frequency distribution.

TABLE 3.6

Natural Logarithm of TTR Data

4.80	4.88	4.99
4.84	4.90	5.05
4.93	4.75	5.00
4.99	4.94	4.74
4.82	5.08	4.82
5.05	4.93	4.92
4.85	4.69	4.80
4.78	4.84	4.98
4.75	4.73	4.71
5.14	4.84	4.87
4.88	4.90	4.88
5.21	5.09	5.25
4.76	5.11	5.15

TABLE 3.7

Natural Logarithm of TTR
Descriptive Statistics

X_{bar}	4.91
s	0.14
Kurtosis	−0.425
Skewness	0.566
n	39
X_{min}	4.69
X_{max}	5.25

4. Calculate the natural logarithm of the TTR data, as shown in Table 3.6.

5. Calculate the descriptive statistics for the logarithmic TTR data, to include the mean and standard deviation, and the sample size, as shown in Table 3.7.

6. Calculate the MTTR in minutes and convert to hours, as shown in Table 3.8. The MTTR is the anti-log of the mean of the natural logarithm of time-to-repair data:

$$\mu_{TTR} = e^{\mu \ln TTR} = e^{4.91} = 136.17 \tag{3.17}$$

TABLE 3.8

Mean TTR

Mean ln (TTR)	4.91
MTTR = $e^{4.91}$ (min)	136.17
136.17/60 (hr)	2.27

TABLE 3.9

Upper Confidence Limit for TTR

$t_{1-\alpha,\nu}$	1.686	$T_{INV}(0.95, 39-1)$
$UCL_{\ln(TTR)}$	5.16	
$UCL_{TTR} = e^{5.16}$ (min)	173.65	
173.65/60 (hr)	2.89	

7. Calculate the upper confidence limit for time to repair in minutes and convert to hours, as shown in Table 3.9. The upper confidence limit is calculated as

$$UCL_{\ln TTR} = \mu_{\ln TTR} + t_{1-\alpha,\nu}\sigma_{\ln TTR} \tag{3.18}$$

where the Student's t sampling distribution is evaluated for $1 - \alpha$ and the degrees of freedom, $\nu = n - 1$. Assuming a 95% confidence limit, the upper confidence limit for the time-to-failure data is 2.89 hours, as shown in Table 3.9.

Weibull Approach

Fitting data to the Weibull repair math model was covered in the failure math modeling section in this chapter. The distinction for repair math modeling is calculating the upper confidence limit. This is readily performed in Excel using Goal Seek. The parameters of the Weibull distribution are found through median ranks regression and correlation analysis, as shown in Table 3.10.

Goal seek finds the value for the upper confidence limit that meets the condition, $F(UCL) = 1 - \alpha$, or C. The procedure is as follows:

1. Enter the value for $1 - \alpha$, or C in a cell.
2. Enter an estimate for the upper confidence limit in a cell. The estimate for the upper confidence limit must be greater than the location parameter.
3. Enter the equation for the Weibull cumulative repair math model in a cell.

Goal seek presents a dialog box that sets the equation for the cumulative repair math model, $F(t)$, to the value for $1 - \alpha$ by changing the value estimated to be the upper confidence limit (Table 3.11).

TABLE 3.10

Weibull Repair Math Model Parameters: Summary Output

Regression Statistics					
r	0.998				
r^2	0.9967				
Adj r^2	0.9966				
SE[a]	0.070				
Obs	39				

ANOVA[b]	**df**	**SS**[c]	**MS**[d]	**F**	**P**	
Regression	1	55.48	55.48	11174	0.000	
Residual	37	0.18	0.00			
Total	38	55.66				

Coefficients		**SE**	**t Stat**	**P**	**LCI**[e] **95%**	**UCI**[f] **95%**
y_0	−5.1235	0.04	−114.76	0.000	−5.21	−5.03
β_3	1.4507	0.01	105.71	0.000	1.42	1.48
η_3	34.18	$e^{-(-5.1235/1.4507)}$				
γ	107					

[a] SE = standard error.
[b] ANOVA = analysis of variance.
[c] SS = sum of squares.
[d] MS = mean of squares.
[e] LCI = lower confidence interval.
[f] UCI = upper confidence interval.

TABLE 3.11

95% Upper Confidence Limit for TTR

Goal seek	
C	0.95
UCL_{TTR} (hr)	179.71 (±3.00)
$F(UCL)$ $(1 − \exp(−(((UCL_{TTR} − \gamma)/\eta)\beta)))$	0.950

Mean Maintenance Time

The mean maintenance time is applied to the following maintainability scenarios:

- Baseline maintenance practice

  Baseline maintenance practices are unscheduled, corrective maintenance actions that are performed to restore a system to functionality.

The baseline maintenance practice may be comprised of two time-to-repair math models: repairs performed in a facility with a controlled environment and repairs performed at the site where the system is located (the two scenarios from the previous example). The baseline mean maintenance time is the weighted average of the time to repair for the two scenarios using the frequency of the scenarios as they weight factors. The equation for the baseline mean maintenance time is expressed in the following equation:

$$\text{MMT}_{\text{Baseline}} = \frac{f_{\text{facility}}\text{MTTR}_{\text{facility}} + f_{\text{location}}\text{MTTR}_{\text{location}}}{f_{\text{facility}} + f_{\text{location}}} \tag{3.19}$$

where f_{facility} is defined as the number of repair events that occur in the facility, $\text{MTTR}_{\text{facility}}$ is defined as the mean time to repair for all maintenance events that occur in the facility, f_{location} is defined as the number of repair of events that occur at the location of the system, and $\text{MTTR}_{\text{location}}$ is defined as the mean time to repair for all maintenance events that occur at the location.

- Preventive maintenance practice

Preventive maintenance practices are scheduled proactive maintenance actions that are performed to preserve system functionality.

$$\text{MMT}_{\text{PM}} = \frac{f_{\text{CT}}\text{MTTR}_{\text{CT}} + f_{\text{PT}}\text{MTTR}_{\text{PT}}}{f_{\text{CT}} + f_{\text{PT}}} \tag{3.20}$$

where f_{CT} is defined as the number of unscheduled maintenance repair events (CT: corrective maintenance time), MTTR_{CT} is defined as the mean time to repair for all unscheduled maintenance repair events, f_{PT} is defined as the number of scheduled maintenance repair events (PT: preventive maintenance time), and MTTR_{PT} is defined as the mean time to repair for all scheduled maintenance events.

Mean Downtime

The baseline mean downtime is the sum of the mean time to repair, the prerepair, and postrepair mean logistics downtime as expressed in the following equation:

$$\text{MDT}_{\text{baseline}} = \text{MTTR}_{\text{baseline}} + L_{\text{prerepair}} + L_{\text{postrepair}} \tag{3.21}$$

Alternatively, the mean downtime can be expressed as the sum of the mean maintenance time and prerepair and postrepair logistics downtime as expressed in the following equation:

$$\text{MDT}_{\text{baseline}} = \text{MMT}_{\text{baseline}} + L_{\text{prerepair}} + L_{\text{postrepair}} \qquad (3.22)$$

The preventive maintenance mean downtime is the sum of mean maintenance time plus the prerepair and postrepair mean logistics downtime as expressed in the following equation:

$$\text{MDT}_{\text{PM}} = \text{MMT}_{\text{PM}} + L_{\text{prerepair}} + L_{\text{postrepair}} \qquad (3.23)$$

System Logistics Downtime Math Model

The logistics downtime model is developed for the system, not a part. Logistics downtime begins with notification that a part failure has occurred causing a system downing event. The logistics events performed to transport the system to the maintenance facility, or transport the maintenance crew to the system, are designated as prerepair logistics downtime. Prerepair logistics downtime also includes all delays that prevent the maintenance action to begin. The delays include waiting for labor, access to the maintenance facility, specialty tools, spare parts, and administrative tasks. The logistics events performed to return a system to operations after the maintenance actions are complete are designated as postrepair logistics downtime. Postrepair logistics downtime includes system test runs, certification of repair actions, transport of the system to the operations location, and administrative tasks (Figure 3.15).

Triangular Distribution

Subjective statistical information fits the triangular distribution. The parameters of the triangular distribution are the minimum, mode, and maximum values for an event. Subjective probability and statistics make use of subject matter experts' (SME) knowledge to evaluate ranges of independent variables such as time. Subject matter experts identify the events required to perform prerepair and postrepair logistics, and they conduct a Delphi survey that specifies the minimum, mode, and maximum time required to perform those events, as shown in Table 3.12. Logistics downtime events are characterized in minutes as with repair model data. The final characterization of the parameters of the triangular distribution will be in hours.

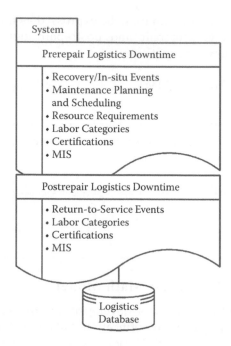

FIGURE 3.15
Qualitative logistics downtime analysis.

Event and Delphi Survey from a Subject Matter Expert

The findings from the Delphi survey for all subject matter experts are used to determine the parameters of the prerepair and postrepair logistics downtime math model as shown in Table 3.13. The estimates for the prerepair and postrepair logistics downtime minimum, mode, and maximum times are the mean of the time estimates provided by the subject matter expert.

The mean logistics downtime for prerepair and postrepair logistics events, Λ, is calculated as the arithmetic mean of the means for the minimum, mode, and maximum times.

The triangular logistics downtime distribution and mean are expressed as follows:

The probability mass density function of the triangular distribution, $f(t)$, consists of four arguments:

- $f(t) = 0$ for all values of time, t, between 0 and T minimum.
- $f(t)$ is fit to a line that begins at T minimum and ascends to the maximum frequency at T mode for all values of time between T minimum and T mode.
- $f(t)$ is fit to a line that begins at T mode and descends to T maximum for all values of time between T mode and T maximum.

TABLE 3.12

Logistical Downtime Events and Subject Matter Expert Delphi Survey Data

	Minutes		
Prerepair logistics events	T_{min}	T_{mode}	T_{max}
Failure notification	15	20	30
Fault detection/isolation	60	75	120
Doc–failure report	15	20	25
Spare part acquisition	45	60	90
Maintenance vehicle/specialty tool queue	45	75	180
Staffing for maintenance action	45	60	90
Sum (min)	225	310	535
λ (min)	356.67		
λ (hr)	5.94		
Postrepair logistics events			
Doc—corrective action	20	30	45
Doc—labor, materials	30	40	60
Doc—repair certification/calibration	20	30	35
Part/materials disposal	30	45	50
Return system to service	0	15	20
Sum (min)	100	160	210
λ (min)	156.67		
λ (hr)	2.61		

TABLE 3.13

Logistics Downtime Math Model Data

Prerepair Logistics Downtime				Postrepair Logistics Downtime			
	T_{min}	T_{mode}	T_{max}		T_{min}	T_{mode}	T_{max}
SME 1	69	90	194	SME 1	35	44	93
SME 2	73	109	176	SME 2	30	66	96
SME 3	66	122	176	SME 3	29	58	102
SME 4	61	111	183	Sum (min)	94	168	291
SME 5	72	95	184	Sum (hr)	1.57	2.80	4.85
Sum (min)	341	527	913	$\lambda_{postrep}$ (hr)			3.07
Sum (hr)	5.68	8.78	15.22				
λ_{prerep} (hr)			9.89				

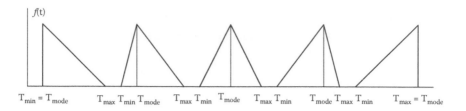

FIGURE 3.16
Range of shapes of the triangular distribution.

- $f(t) = 0$ for all values of time greater than T maximum.

$$
f(t) = \begin{vmatrix}
0 & \text{if} & 0 \le t < T_{min} \\
\dfrac{2(t - T_{min})}{(T_{max} - T_{Min})(T_{mode} - T_{min})} & \text{if} & T_{min} \le t < T_{mode} \\
\dfrac{2(T_{max} - t)}{(T_{max} - T_{Min})(T_{max} - T_{mode})} & \text{if} & T_{mode} \le t \le T_{max} \\
0 & \text{if} & t > T_{max}
\end{vmatrix}
\tag{3.24}
$$

The mean of the triangular distribution is the sum of the three values for time divided by three as shown in the following equation.

$$
\mu_\Lambda = \frac{T_{min} + T_{mode} + T_{max}}{3}
\tag{3.25}
$$

The shape of the triangular distribution mass density function can range from a right triangle skewed to the right, a positively skewed distribution; asymmetrical distribution about the mean, a negatively skewed distribution; and a right triangle skewed to the left, as shown in Figure 3.16.

Summary of Reliability Math Models

Reliability math models for time to failure, prerepair logistics downtime, time to repair, and postrepair logistics downtime describe the operational and maintenance cycle of a part over the useful life of a system. The math models are presented in a timeline for notional shapes in Figure 3.17. Time-to-failure math model describes the expectation of the incidence of part failure. Prerepair logistics downtime models describe the expectation for the duration of prerepair logistics events immediately following part failure.

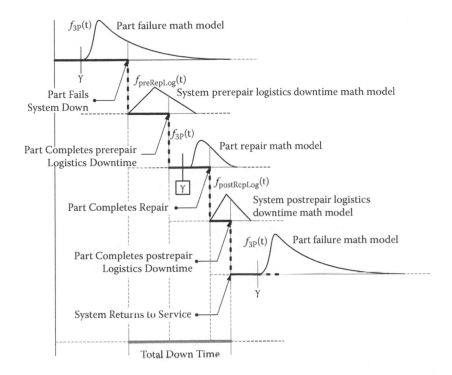

FIGURE 3.17
Part failure maintenance cycle reliability math models.

Time-to-repair math models describe the expectation of the duration of maintenance events. Postrepair logistics downtime models describe the expectation for the duration of postrepair logistics events immediately following the completion of maintenance actions.

Availability

Availability is the probability that a part, an assembly, up to the system design configuration will be able to perform its mission when required. The basic equation for availability is uptime divided by the sum of uptime and downtime. Availability for higher-level design configurations is calculated from part availability in the same way that reliability of higher-level configurations is calculated from part reliability. Design engineers are called upon to calculate inherent and operational availability for parts. The distinction between inherent and operational availability is based on the definition of uptime and downtime.

Operational Availability, A_o

Operational availability is a predictive metric calculated for the baseline reliability analysis. The calculation for operational availability for a part is expressed in the following equation:

$$A_O = \frac{\text{MTBF}}{\text{MTBF} + \text{MMT}_{\text{baseline}} + \Lambda_{\text{prerepair}} + \Lambda_{\text{postrepair}}} \qquad (3.26)$$

where the baseline mean maintenance time is used to reflect the baseline corrective maintenance practice that includes performance of maintenance actions in both the facility and the system location.

Uptime for operational availability is defined as the part mean time between failure (MTBF). Downtime for the operational availability is defined as the sum of the baseline mean maintenance time (MMT$_{\text{baseline}}$), and the mean logistics downtime (MLDT). The mean logistics downtime is equal to the sum of the mean prerepair logistics downtime, $\Lambda_{\text{prerepair}}$, and the mean postrepair logistics downtime, $\Lambda_{\text{postrepair}}$.

Achieved Availability, A_a

Achieved availability is a maintenance management metric calculated for the revised reliability analysis that reflects the change in availability resulting from the incremental continuous improvement for maintenance and logistics practices. The calculation for achieved availability for a part is expressed in the following equation:

$$A_a = \frac{\text{MTBF}}{\text{MTBF} + \text{MMT}_{\text{PM}} + \Lambda_{\text{prerepair}} + \Lambda_{\text{postrepair}}} \qquad (3.27)$$

or

$$A_a = \frac{\text{MTBF}}{\text{MTBF} + \text{MDT}} \qquad (3.28)$$

Application of Reliability Math Models

The development of the reliability math models for parts provides the information that is used to develop reliability simulations that use the distributions in a Monte Carlo random variable generator to characterize the

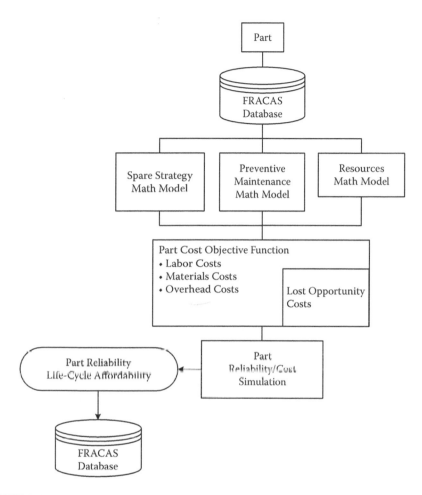

FIGURE 3.18
Quantitative reliability analyses inputs to reliability simulation.

operational and maintenance cycles over the useful life of the asset for critical parts. The reliability simulations characterize the behavior of the part operational and maintenance cycles to describe the effects on the reliability, maintainability, and availability of the critical part. The logic of reliability simulation is provided in Figure 3.18. Reliability simulation is addressed in Chapter 9, Reliability Simulation and Analysis.

FIGURE 5.1

4

Engineering Economic Analysis

The objective of engineering economic analysis is to investigate and evaluate the costs of two or more decision alternatives by application of accounting and financial methods that calculate present, future, and recurring amounts and equivalent present and uniform values.

The accounting and financial methods that constitute engineering economic analysis that will be applied to reliability-based life-cycle economic analysis are summarized in this chapter. Detailed examples of computations are provided in Appendix D: Engineering Economic Analysis.

The distinctions between business accounting and finance methods performed by an organization and engineering economic analysis are defined by the limited application to evaluate decision alternatives. Engineering economic analyses focus on specific direct costs associated with the decision alternatives and rarely include revenue. Reliability decision alternatives are typically defined as follows:

- Investments in performing reliability analysis for a part versus the do-nothing alternative
- Trade studies for two or more reliability options

The role of the manager is to identify and build a team of multidiscipline professionals in finance, accounting, and engineering who possess all of the information necessary to perform an engineering economic analysis. The role of the engineer is to perform the reliability analyses that provide the demand function for resources.

Engineering Economic Analysis Information

Engineering economic analysis information is an understanding of cash flow. Cash flow is determined by the sources and uses of cash. Sources of cash can be categorized as long-term capital and short-term capital, also referred to as operating capital. Long-term capital finances the acquisition of capital assets and consists of equity, bonds, and debt. The cost to service long-term capital contributes to the calculation of an organization's required internal rate of return, also referred to as the discount rate. The evaluation of decision alternatives will rely heavily on the discount rate.

Short-term or operating capital finances day-to-day operations of an organization and consists of lines of credit, corporate paper, purchase orders, and revenue. The evaluation of decision alternatives investigates the consumption of operating capital that is used to pay labor, materials, and overhead expenses. Lines of credit and corporate paper carry an interest expense and require repayment of the principal balance. Purchase orders are equivalent to short-term loans provided by the vendor of products and services. Although no interest rate is specified, the vendor will calculate the price of the product or service to include a provision for their discount rate over the term of the purchase order. Revenue that is reinvested in the organization is expected to yield a return equal to or greater than an organization's discount rate.

Uses of cash can be categorized as capital costs and operating and maintenance costs. Capital costs serviced through the payment of interest expenses, capital recovery, and sinking funds are included in operating and maintenance costs. Interest expenses are the cost of money levied on the organization by the lender and are expressed in annual percentage rate (APR). Capital recovery is a loan payment schedule that includes repayment of interest and pay down of the outstanding loan balance, i.e., a mortgage or car payment loan. Sinking fund is a loan payment schedule that includes repayment of interest and a separate income earning investment that will grow to the amount due at the end of the loan term, i.e., term loans or bonds.

Engineering economic analysis limits its investigation of capital costs and the servicing expenses for direct investment in a reliability program. Capital costs associated with an organization's infrastructure are not included in the engineering economic analysis except that the servicing costs determine the organization's discount rate. For example, an organization that has an existing maintenance facility accounts for the capital costs associated with its construction. A reliability investigation that determines a requirement for a modification to the facility that is funded by capital investment will include the capital cost and its servicing in its economic analysis. The existing capital cost and its servicing expenses are considered to be sunk costs in that they will continue to exist regardless of the decision made by the engineering economic analysis, and will not be included.

Operating and maintenance costs consist of direct labor, direct materials, and direct overhead expenses. Operating costs are associated with revenue production. Maintenance costs are associated with system sustainment. Direct labor expenses are the costs to the organization incurred through the employment of people. Direct labor expenses include the employee's compensation, direct overhead, and indirect overhead. Compensation includes the salary or wage paid to the employee plus the cost of fringe benefits, government fees, and incentive payments. Direct overhead includes the costs of supporting infrastructure, including first-line supervisor, support and administrative staff, departmental facilities and equipment, and training.

Indirect overhead includes all of the remaining costs of operating the business, including corporate headquarters and staff, finance and accounting departments, and marketing and sales.

Fundamental Economic Concepts

Engineering economic analysis is the application of the following fundamental economic concepts.

1. Time value of money

   The time value of money recognizes that a dollar today is not the same as a dollar next year. It is essential that each cost of event must be identified by its magnitude and when it occurs.

2. Internal rate of return, IRR, r

   The internal rate of return is expressed as a percentage, r, and describes two metrics. The first metric is a complex determination of the cost of money for an organization for all sources of money. An organization will calculate this internal rate of return, or discount rate, periodically. The discount rate is used to evaluate all investment alternatives. The typical decision rule is that an investment alternative must yield a return on investment that is at least equal to, but preferably greater than, the discount rate. It is not enough that an investment alternative is profitable.

   The second metric is the internal rate of return, IRR, that is expected from an investment alternative, also referred to as the return on investment. This annualized internal rate of return is calculated as nth root of the product of the quantities $(1 + i_t) - 1$, where i_t is the rate of return for the tth year and is expressed as

   $$\text{IRR} = \sqrt[n]{\prod_{t=0}^{n}(1+i_t)} - 1 \tag{4.1}$$

   where $t = 0$ is the beginning of the evaluation period, $t = n$ is the current year, and i_t is the rate of return for each year.

3. Duration, N

   The duration of a project is expressed in years. For example, a project's duration of 18 months is expressed as $N = 1.5$ (18 months/12 months/year). Engineering economic analyses are typically applied to projects that have fixed durations. The duration of reliability investment alternatives are typically characterized by the useful life of a system, or a part.

4. Compounding periods per year, m

   Compounding periods per year are defined by the frequency of cost events, to include weekly, monthly, quarterly, semiannually, and yearly, $m = 52, 12, 4, 2$, and 1, respectively. The number of compounding periods per year is influenced by the duration of the project and an organization's practices. For $N \leq 5$ yr, the selection for m is typically monthly. For $N > 5$ yr, the minimum selection for m is quarterly; typically annually. The deciding factor for the number of compounding periods per year is based on judgment; finding the frequency that makes the most sense to the organization.

5. Number of compounding periods, n

   The number of compounding periods in a project is calculated as the product of the duration and compounding periods per year, $n = mN$.

6. Interest rate per compounding period, i

   The internal rate of return per compounding period is calculated as the internal rate of return divided by the number of compounding periods per year, r/m.

Amounts versus Equivalent Value

Cash events that occur over time are defined as "amounts." Equivalent cash events are defined as "values." For example, direct labor, materials, and overhead expenses allocated in the time period in which they occur are "amounts." The equivalent direct labor, materials, and overhead expenses calculated at a specific point in time are "values."

Cost Estimation

Cost estimation begins with the calculation of all categories of expenses that exist in the current time period, at $t = 0$. Accuracy of cost estimation in the current time period is crucial to an engineering economic analysis. The expectation that these estimates are performed with certainty is overshadowed by the need for accuracy.

Engineering economic analysis of reliability investigations is driven by the same project management approach used for any other project. A work breakdown structure for tasks and milestones must be performed prior to cost estimation. Each task will be assigned employees from various crafts and grades, supervision and management, and administration. Each employee labor category will have a designated cost per hour that will be allocated to when their work is performed within the task. The number of employees by labor category will be assigned to each task. Materials and direct overhead expenses will also be assigned by task and when they occur. The only source of certainty that exists in cost estimation for the initial time period is the hourly cost for each employee. Labor allocation, labor numbers, materials, and direct overhead expenses are uncertain estimates.

Cost estimation of all categories of expenses that occur in the future are determined from cost escalators. Cost escalators for labor and materials will typically be different. Cost escalators for various categories of labor will also be different. The frequency of cost escalators for labor will vary by employee category. Cost escalators for materials will vary by material with a worst-case that every material has its own cost escalator. Cost escalators for labor and materials vary from year to year, adding to the uncertainty of future cost estimation. Fortunately, engineering economic analysis is not invalidated by changes in the future cost escalators since the error between the used escalator and the actual escalator cancel out. But future cost estimation is worthless if current cost estimation is not accurate!

Classifications of Sources and Uses of Estimated Cash

Cash flow uses the convention where sources of cash and credits are positive and uses of cash are negative.

1. Present amount, P

   The present amount is an estimated cost that occurs in the initial period, $t = 0$.

2. Future amount, F, F_t

   The future amount, F_t, is an estimated cost that occurs in the periods following the current period, $t = 0$. The future amount may be known with certainty if it is specified by an agreement or estimation. For example, a labor contract will specify the labor cost escalator over the duration of a contract. A vendor of a system recommends an overhaul at 3000 engine hours with an expected cost at that time.

   The future amount for costs that vary with time (labor, materials, and overhead) is calculated from the present amount using a cost escalator, g. For example, the current labor cost per hour is $10. The labor cost escalator is estimated to be 5%. The future amount in the next year, F_1, is expected to be $10(1 + g) = $10(1.05) = 10.50 per hour. The general expression for future amounts of a present amount is

$$F_t = P(1+g)^t \qquad (4.2)$$

where t is defined as the compounding period in which the future amount occurs.

Cost escalators for labor and materials can be found in Department of Commerce publications and their web page. Labor cost escalators are typically the consumer price index or the inflation rate. Local chambers of commerce publish the consumer price index that applies to that region. There are as many cost escalators for materials as there are types of materials. A generalized cost escalator for materials is the wholesale price index.

And then there are the future amounts that defy analysis. Technological innovation wreaks havoc on analytical approaches to cost estimation based on cost escalators or trended data. Prices for innovations typically deescalate as the innovation matures. Commodity prices, particularly for energy, defy estimation. Commodity prices go up and down over time. Obsolescence is a barrier to estimating future amounts. Cost escalators for such are estimated from engineering and management judgment.

[Note that error in the respective future amounts for two or more projects cancels out if the actual cost escalator next year is different from the current estimate.]

3. Recurring amounts, A, A_t

   Recurring amounts are costs that occur on a fixed frequency, weekly, monthly, etc. The magnitude of a recurring amount in time period t is designated as A_t. All recurring amounts are future amounts. Recurring amounts are either uniform or increasing. Uniform recurring amounts remain constant over time:

$$A \mid A_1 = A_2 = \ldots = A_t \qquad (4.3)$$

Examples include interest payments on a term loan, payments to a sinking fund to retire a loan, and capital recovery payments for a loan. Labor, materials, and overhead expenses that vary over time may behave like a uniform recurring amount for project durations less than a year.

Increasing recurring amounts are either linear or geometric. Linear increasing recurring amounts, A_t, grow at a constant magnitude, G, over time. The initial magnitude of a linear increasing recurring amount occurs at time period $t = 1$ and is designated as A_1. The magnitude of the next linear increasing recurring amount is

$$A_2 = A_1 + G \tag{4.4}$$

The magnitude of the third linear increasing recurring amount is

$$A_3 = A_2 + G = A_1 + G + G = A_1 + 2G \tag{4.5}$$

The general expression for the magnitudes of the linear increasing recurring amount is expressed as

$$A_t = A_1 + (t - 1)G \tag{4.6}$$

Geometric increasing recurring amounts, A_t, grow at a constant rate, g, over time. The initial magnitude of a geometric increasing recurring amount occurs at time period $t = 1$ and is designated as A_1. The magnitude of the next linear increasing recurring amount is

$$A_2 = A_1(1 + g) \tag{4.7}$$

The magnitude of the third linear increasing recurring amount is

$$A_3 = A_2(1 + g) = A_1(1 + g)(1 + g) = A_1(1 + g)^2 \tag{4.8}$$

The general expression for the magnitudes of the geometric increasing recurring amount is expressed as

$$A_t = A_1(1 + g)^{t-1} \tag{4.9}$$

Geometric increasing recurring amounts describe cost escalators applied to labor, materials, and overhead (Figure 4.1).

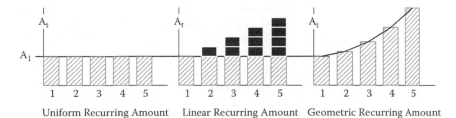

FIGURE 4.1
Recurring amounts.

Cash Flow Timeline

The objective of developing a cash flow timeline is to calculate the net cash flow (NCF) per compounding period for the project. The net cash flow per compounding period is the sum of all cost events that occurred during that compounding period.

Spreadsheets are an ideal approach to performing engineering economic analysis. All of the cost estimation information is fit to a cash flow timeline on the spreadsheet. Much of the cost estimation for future amounts can be calculated in the spreadsheet. The construction of a cash flow timeline consists of row designations for the timeline units, days, weeks, months, etc., and column designations for all cost categories. Timeline units defined by compounding periods, days, weeks, months, etc., are the preferred choice. Timeline units defined by calendar units that differ from compounding periods complicate the calculations for future amounts and equivalent values. For example, a timeline using weeks for compounding periods measured in months will require intermediate steps in computing future amounts and equivalent values. An example of a cash flow timeline is provided in Table 4.1.

Cost categories include capital cost expenses (interest expenses, salvage value, capital recovery expenses, sinking fund expenses, and depreciation) and operating and maintenance expenses (direct labor by labor category, materials by category, and direct overhead by category).

Equivalent Values

Equivalent values allow the comparison of cost events over time for two or more projects. The net present value of a project can be used to compare two or more projects if each project occurs over the same duration. For example, the net present value of projects that occur for five years each can be compared by their net present values. The equivalent uniform value for a project can be used to compare two or more projects when each project occurs over a different duration.

The procedure to calculate the net present value of a project is to calculate the present value of the net cash flow for each reporting period and add them.

The procedure to calculate the equivalent uniform value is to calculate the net present value for each project and then calculate the uniform value for each net present value.

TABLE 4.1

Cost Categories

Cash Flow Timeline: Retrieval Vehicle

	Capital Costs ($)				Sinking Fund ($)		Capital Tax ($)		
	Loan	Cap Acq	PMT_{Loan}	SV	PMT_{SF}	Bal_{SF}	PMT_{Dep}	BookVal	PMT_{CT}
t	(2)	(3)	(4)	(5)	(6)	(7)	(8)	(9)	(10)
(1)									
0	28,000	−28,000						28,000	
1					−688	688	−314	27,686	−50
2					−688	1,381	−314	27,373	−50
3			−315		−688	2,079	−314	27,059	−49

	Operating and Maintenance Costs ($)				Tax Table ($)				NCF ($)	
	Exp_{DLO}	Exp_{DLM}	Exp_{Matls}	$Exp_{DO/H}$	ΣFTD_{Exp}	F_{Tax}	ΣSTD_{Exp}	S_{Tax}	Pretax	Posttax
t	(11)	(12)	(13)	(14)	(15)	(16)	(17)	(18)	(19)	(20)
(1)										
0										
1	−34,253	−4,757	−1,111	−2,716	−43,201	16,848	−26,353	1,739	−43,525	−24,988
2	−34,253	−4,757	−1,116	−2,716	−43,206	16,850	−26,355	1,739	−43,530	−24,990
3	−34,253	−4,757	−1,121	−2,716	−43,525	16,975	−26,550	1,752	−43,851	−25,172

1. Equivalent present value, PV

   The equivalent present value, PV, can be calculated for a future amount, F, and a uniform recurring amount, A, when the following information is known:

   - r, internal rate of return
   - N, project duration in years
   - m, compounding periods per year

   The equivalent present value of a future amount is expressed in the following equation:

   $$PV = \frac{F}{(1+i)^n} \qquad (4.10)$$

   where i is defined as the interest rate per compounding period, IRR/m, and n is defined as the total number of compounding periods in the project, mN.

   MS Excel™ provides financial analysis functions.[*] The Excel financial function dialog box for calculating present value is

   PV(rate, nper, pmt, fv)

   where

   > rate is the cell designation for interest rate per compounding period, i
   >
   > nper is the cell designation for number of compounding periods, n
   >
   > pmt is the cell designation for uniform recurring amount, A
   >
   > fv is the cell designation for future amount, F

   The PV function calculates the present value for either a uniform recurring amount or a future amount. The pmt entry is left blank when calculating the present value of a future amount, e.g.,

   PV(rate, nper,, fv).

---

[*] MS Excel™ financial functions have a quirk—the result for a financial function changes the sign; if the input amount is negative the output amount will be positive. The engineer must be careful to assure that the output has the appropriate sign. One way is to apply the negative sign to the financial function, i.e., –FV, –PMT, and –FV.

The equivalent present value of a uniform recurring amount is expressed in the following equation:

$$PV = A\left[\frac{(1+i)^n - 1}{i(1+i)^n}\right] \tag{4.11}$$

The Excel function for calculating the present value of a uniform recurring amount is expressed as

PV(rate, nper, pmt)

The net present value, NPV, is the sum of the present values for each compounding period, PV_t, expressed in the following equation:

$$NPV = \sum_{t=0}^{n} PV_t \tag{4.12}$$

The Excel function for calculating the net present value of a column of net cash flows is expressed as

NPV(rate, $NCF_{t=1}$:$NCF_{t=n}$)

where rate is the internal rate of return per compounding period, and $NCF_{t=1}$:$NCF_{t=n}$ is the range of net cash flows from the first compounding period through the last, nth, period. The net cash flow when a value occurs in the present period, $t = 0$, is added to the Excel net present value function, not included in the range of NCF values.

2. Equivalent uniform value, A_{equv}

The equivalent uniform value, A_{equv}, is calculated for the net present value, NPV, and is expressed in the following equation.

$$A_{equv} = NPV\left[\frac{i(1+i)^n}{(1+i)^n - 1}\right] \tag{4.13}$$

The Excel function for calculating the equivalent uniform value, A_{equv}, for the net present value is expressed as

PMT(rate, nper, pv)

where rate is the IRR per compounding period.

Example 1: Project Cash Flow Timeline: m = 1

The project duration is scheduled to be three years with annual compounding periods. The organization's discount rate, IRR, is provided by management, $r_{IRR} = 9\%$. The project will require financing. Two sources of debt are acquired: a term loan for capital acquisition of machinery and a capital recovery loan for acquisition of a vehicle. The project will incur operating expenses for labor, materials, and overhead.

Cash Events

Term Loan: $10,000 at Simple Interest of $r_{TL} = 6\%$ APR

The term loan provides three cash events:

1. proceeds of the loan at $t = 0$
2. repayment of the loan at $t = 3$
3. interest payments on the loan at $t = 1, 2,$ and 3

The proceeds of the term loan are a present amount $P_{TL} = -\$10,000$. The proceeds from the term loan will be received and the money committed in period 0 for a net cash flow of $0. The term loan must be paid in full at the end of the third year, which is a future amount, $F_{TL} = \$10,000$. Simple interest will be paid for the term loan at the end of each of the three years, which is a uniform recurring amount, $A = i_t = r_{TL}P_{TL} = -\$600/\text{yr}$.

Sinking Fund: $10,000 at Compound Interest of $r_{SF} = 7\%$ APR

A sinking fund will be established by investing revenue from operations in a money market fund to assure that the organization will be able to pay the term loan at the end of the third year. The term loan provides two cash events:

1. investment at $t = 1, 2,$ and 3
2. revenue from principal and earnings at $t = 3$

The sinking fund investments at the end of each year are a uniform recurring amount, $A = -\$3,111/\text{yr}$, calculated in Excel using the PMT function:

$$\text{PMT}(7\%, 3, \_, -\$10,000) = -\$3,111/\text{yr}$$

where 7% is the interest rate per compounding period for $m = 1$, 3 is the total number of compounding periods, the space is the blank placeholder for present value, and −$10,000, is the future amount the sinking fund must achieve to pay the term loan. In year 3 the value of the sinking fund will be used to pay the balance of the term loan resulting in a net cash flow of $0.

The revenue from the sinking fund is a future amount, F_{SF} = $10,000. The future amount of a sinking fund is assumed to equal the present amount of the term loan, $F_{SF} = P_{TL}$ = $10,000, with a change in sign.

[Note: the sum of the sinking fund payments, $3,111, over three years is less than $10,000, but the future amount of the sinking fund grows to equals $10,000 due to the money market interest earnings.]

[Another note for sinking funds: the frequency of compounding periods need not be equal for term loans the respective sinking fund.]

Capital Recovery Loan: $12,000 at Compound Interest of r_{CR} = 5% APR

The capital recovery loan provides two cash events:

1. proceeds of the loan at $t = 0$
2. loan payments at $t = 1, 2$, and 3

The proceeds of the capital recovery loan are a present amount P_{CR} = −$12,000. The proceeds from the capital recovery loan will be received and the money committed in period 0 for a net cash flow of $0. Equal annual loan payment will occur at t − 1, 2, and 3, which is a uniform recurring amount, A = −$4,407/yr. The uniform recurring payment on the capital recovery loan is calculated in Excel using the PMT function:

$$PMT(5\%, 3, \$12,000) = -\$4,407/yr$$

where 5% is the interest per compounding period for m = 1, 3 is the total number of compounding periods, mN, and $12,000 is the present amount at $t = 0$.

[Note that the sum of $4,407 over three years is greater than $12,000. The annual payments include interest per year on the unpaid loan balance plus payments on the loan principal.]

Operating Expenses

Operating expenses provide three cash events:

1. Labor expenses at $t = 1, 2$, and 3
2. Materials expenses at $t = 1, 2$, and 3
3. Overhead expenses at $t = 1, 2$, and 3

While each cost event is a recurring event, each increases geometrically over the project term. If the cost escalation factor was equal for each cost event, the three could be summed as a single recurring amount. But that is not the case. Cost estimations for labor, materials, and overhead for the first project period are $A_1 = -\$12,000$, $-\$9,000$, and $-\$3,000/\text{yr}$ respectively. Each is a geometrically increasing recurring cost with a cost escalator, e.g., labor costs are expected to grow by $g_L = 5\%$ per year, materials by $g_M = 3\%/\text{yr}$, and overhead by $g_{OH} = 2\%/\text{yr}$. The general equation for geometrical increasing recurring amounts is

$$L_t = L_{t-1}(1 + g) \tag{4.14}$$

The labor cost in period 2, L_2, is estimated by

$$L_2 = L_1(1 + g_L) = -\$12,000(1.05) = -\$12,600 \tag{4.15}$$

The material cost in period 2, M_2, is estimated by

$$M_2 = M_1(1 + g_M) = -\$9,000(1.03) = -\$9,270 \tag{4.16}$$

The overhead cost in period 2, OH_2, is estimated by

$$OH_2 = OH_1(1 + g_{OH}) = -\$3,000(1.02) = -\$3,060 \tag{4.17}$$

The economic information for project 1 is tabulated in Excel™ in Table 4.2.

The cash flow timeline is completed for all costs categories for the time range of $t = 0$ to $t = n$ in Table 4.3. The net cash flow per compounding period, NCF_t, is calculated as the sum of all cash events, i, $C_{i,t}$, for each time period, t,

$$NCF_t = \Sigma(C_{i,t}) \tag{4.18}$$

TABLE 4.2

Economic Information
for Project 1 for $m = 1$

r_{IRR} (%)	9
P_{TL} ($)	10,000
r_{TL} (%)	6
r_{SF} (%)	7
P_{CR} ($)	12,000
r_{CR} (%)	5
N	3
m	1
i_{IRR} (%)	9
n	3
g_L (%)	5
g_M (%)	3
g_{OH} (%)	2

TABLE 4.3

Cash Flow Timeline for $m = 1$

Cash Flow Timeline

t	Capital Costs ($)			O&M Costs ($)			NCF$_t$ ($)	PV$_t$ ($)
	i_{TL}	SF	CR	Labor	Materials	Overhead		
0	0	0	0	0	0	0	0	0
1	-600	-3,111	-4,407	-12,000	-9,000	-3,000	-32,117	-29,465
2	-600	-3,111	-4,407	-12,600	-9,270	-3,060	-33,047	-27,815
3	-600	-3,111	-4,407	-13,892	-9,835	-3,184	-35,027	-27,047
$\Sigma =$	-1,800	-9,332	-13,220	-38,492	-28,105	-9,244	-100,191	-84,327

$= -PV(i_{IRR}, n, \neg, NCF_t)$

$-84,327 = \Sigma(PV_t) = NPV$

$FV_{SF} = \$10,000 = FV(i_{IRR}, n, SF)$

$NPV(r, NCR_t = 1; NCF_t = n) = -\$84,327$ $\Sigma(NCF_t)$

$i_{TL} = r_{TL}P_{TL}$

$SF = PMT(r_{SF}, n, \neg, P_{TL})$

$CR = PMT(r_{CR}, n, \neg, P_{CR})$

$L_t = L_t - 1(1 + g_L)t - 1$

$M_t = M_t - 1(1 + g_M)t - 1$

$OH_t = OH_t - 1(1 + g_{OH})t - 1$

$A_{equiv} = -\$33,314 = -PMT(i_{IRR}, n, NPV)$

The present value for each time period is calculated by the Excel PV function,

$$PV(rate, nper, \_, fv)$$

where rate $= i_{IRR}$, nper $= t$, pmt is left blank, and fv $= NCF_t$,
The present value for the net cash flow that occurs in period 1 is

$$PV(0.09, 1,, -\$32,117) = -\$29,465$$

where 0.09 is the discount rate per time period for $m = 1$, 1 is the time period, two commas indicate a blank placeholder for uniform payments, and $-\$32,117$ is the future amount that occurs in time period 1. In other words, the present value calculation per time period treats each net cash flow as a future amount that occurs in that time period.

The net present value is the sum of all present values from $t = 0$ to $t = n$:

$$NPV = \Sigma(PV_t) = -\$84,327 \qquad (4.19)$$

The Excel net present value function, NPV, is calculated as

$$NPV(rate, NCF_{t=1}:NCF_{t=n})$$

where rate is the discount rate per time period and $NCF_{t=1}:NCF_{t=n}$ is the range of net cash flows per time period from $t = 1$ to $t = n$. When a present amount exists for $t = 0$, then it is added to the Excel NPV calculation:

$$NPV(rate, NCF_{t=1}:NCF_{t=n}) + P_{t=0}$$

The net present value for the series of net cash flows in this example is

$$NPV(0.09, -\$32,117:-\$35,027) = -\$84,327$$

where 0.09 is the discount rate per compounding periods for $m = 1$, and $-\$32,117:-\$35,027$ is the range of net cash flows from $t = 1$ to $t = 3$.

The equivalent uniform amount is found by the Excel PMT function for the net present value,

$$PMT(rate, nper, pv)$$

where rate $= i_{IRR}$, nper $= n$, and pv $= NPV$.
The equivalent uniform amount for this example is

$$PMT(0.09, 3, -\$84,327) = -\$33,314$$

where 0.09 is the discount rate for $m = 1$, 3 is the total number of discount periods, and $-\$84,327$ is the net present value of the project.

Cash Flow Timeline

The completed cash flow timeline, including uniform recurring capital costs (interest payments for the term loan, i_t; investments in the sink fund, SF; payments on the capital recovery loan, CR); recurring operating and maintenance costs (labor, materials, overhead); calculation of the net cash flow per compounding period, NCF_t; present value for each time period, PV_t; net present value of the net cash flows, NPV; net present value of the present values for each time period, $\Sigma(PV_t)$; and the equivalent uniform amount, A_{equv}, is provided in Table 4.3.

At this point the reader might ask, "So What?"

The following observations can be made:

- The sum of the series of cash flows (−$100,191) over three years is equivalent to −$84,327 in the current time period, $t = 0$.
- The series of cash flows (−$100,191) over three years is equivalent to a uniform recurring amount of −$33,314/yr over three years.
- In engineering economic terms, given the internal rate of return, an organization is indifferent to the series of net cash flows over three years, or the equivalent net present value in the current time period, $t = 0$, or the equivalent uniform recurring amount over three years.

But, all three observations are meaningless in the absence of alternatives.

Assume a second alternative project, P_2, is investigated with the following information:

$$N_2 = 3, \text{NPV}_2 = -\$92,000, \text{ and } A_{equv2} = -\$45,000/yr$$

When the project durations are equal, $N_1 = N_2$, the decision criterion is to select the lowest NPV: Project 1 evaluated above, P_1, −$84,327, is the lower equivalent cost compared to −$92,000 for P_2, $\text{NPV}_1 < \text{NPV}_2$.

Assume a third alternative project, P_3, is investigated with the following information:

$$N_3 = 5, \text{NPV}_3 = -\$80,000, \text{ and } A_{equv3} = -\$36,000/yr$$

When the project durations are not equal, $N_1 < N_3$, the decision criterion is to select the lowest equivalent recurring amount: A_{equv} project P_1, −$33,314/yr is the lower cost compared to −$45,000/yr for P_2 and −$36,000/yr for P_3 even though P_3 has the lowest NPV = −$80,000.

Example 2: Project Cash Flow Timeline: $m = 4$

Let example 1 be modified by changing the compounding periods per year from 1 (annually) to 4 (quarterly). The revised calculations for the internal

TABLE 4.4

Economic Information
for Project 1 for $m = 4$

r_{IRR} (%)	9
P_{TL} ($)	10,000
r_{TL} (%)	6
i_{TL}	0.015
r_{SF} (%)	7
i_{SF}	0.018
P_{CR} ($)	12,000
r_{CR} (%)	5
i_{CR}	0.013
N	3
m	4
i_{IRR}	0.023
n	12
g_L (%)	5
g_M (%)	3
g_{OH} (%)	2

rate of return and loan interest rates per compounding period as well as the number of compounding periods are provided in Table 4.4. The quarterly interest rate for the term loan is calculated as

$$i_{TL} = r_{TL}/m = 6\%/4 = 0.015$$

The quarterly interest rate for the sinking fund is calculated as

$$i_{SF} = r_{SF}/m = 7\%/4 = 0.018$$

The quarterly interest rate for the capital recovery loan is calculated as

$$i_{CR} = r_{CR}/m = 5\%/4 = 0.013$$

The number of compounding periods is calculated as

$$n = mN = (4)(3) = 12$$

The economic information for project 2 is tabulated in Excel in Table 4.4.

The cash flow timeline row designation for compounding periods, t, is changed to reflect the number of quarters in the evaluation period, $t = 0, 1, 2, \ldots, 12$. The simple interest on the term loan is calculated quarterly

$$i_t = i_{TL}P_{TL} = (0.015)(-\$10{,}000) = -\$150/\text{qtr}$$

The sinking fund payment is calculated as

$$\text{PMT}(0.018, 12, \_, -\$10{,}000) = -\$756/\text{qtr}$$

The capital recovery loan payment is calculated as

$$\text{PMT}(0.013, 12, \_, -\$12{,}000) = -\$1{,}083/\text{qtr}$$

The quarterly expenses for labor, materials, and overhead for the first year are calculated by dividing the annual costs from example 1 by the number of compounding periods in the year, $m = 4$. But, the escalation factors are applied annually. In this way, escalation factors do not behave like interest rates. For example, the labor expense is $-\$3{,}000$/quarter in the first year for all four quarters. The escalation factor applies to the first compounding period in the second year. The frequency of labor, materials, and overhead escalation factors often differ from each other. Labor escalation factors for hourly employees might be applied semiannually, while salaried employees are typically applied annually. Materials escalation factors might be applied quarterly. As previously mentioned, there may be as many escalation factors as there are materials categories. Overhead escalation factors might be applied every two years.

The net cash flow for each compounding period is equal to the sum of the cost events for each compounding period.

The present value for each time period is calculated by the Excel present value function:

$$\text{PV}(i, n, \_, \text{NCF}_t)$$

The present value for the first quarter is calculated as

$$\text{PV}(0.023, 1, \_, -\$7{,}989) = -\$7{,}813.$$

The net present value for the project cash flows is calculated as

$$\text{NPV} = \sum_{t=0}^{n} \text{PV}_t \tag{4.20}$$

The net present value for the project cash flows is equal to $-\$85{,}515$. The net present value for the project cash flows can also be calculated using the Excel net present value function

$$\text{NPV}(i, \text{NCF}_t = 1{:}\text{NCF}_{t=n})$$

The net present value for the series of net cash flows in this example is

$$\text{NPV}(0.023, -\$7,989: -\$8,464) = -\$85,815.$$

The equivalent uniform amount is calculated by the Excel PMT function for the net present value

$$\text{PMT(rate, nper, pv)}$$

where rate $= i_{\text{IRR}}$, nper $= n$, and pv $= \text{NPV}$.
  The equivalent uniform amount for this example is

$$\text{PMT}(0.023, 12, -\$85,815) = -\$8,211.$$

The completed cash flow timeline, including the calculation of the present value for each time period, the net present value of the net cash flows, and the equivalent uniform amount, is provided in Table 4.5.
  [NOTE: Escalation factors are applied annually.]

Example 3: Project Cash Flow Timelines for Three Projects: $m = 1$

Three alternatives for a reliability investigation are investigated, P_1, P_2, and P_3, and the respective net cash flows are summarized in Table 4.6. Each project's net cash flow sums to the same amount, $1750. The net present value is evaluated at an internal rate of return equal to 6%. Project 3 is selected because it has the minimum net present value for cost. The net present value for cost is the decision discriminator because the durations of the three projects are identical. Although not necessary, the equivalent uniform amount for each project confirms the selection of project three based on the net present value.

Example 4: Project Cash Flow Timelines for Three Projects: $m = 1$

A reliability analysis recommends the acquisition of a specialty tool to be used for system sustainment. The duration of the need for the specialty tool is indefinite. Three alternatives are proposed, Projects 4, 5, and 6. An

TABLE 4.5

Cash Flow Diagram Timeline for Project 1 for $m = 4$

	Capital Costs ($)			O&M Costs ($)			NCF_i ($)	PV_i ($)
	i_{TL}	SF	CR	Labor	Materials	Overhead		
0	0	0	0	0	0	0	0	0
1	−150	−756	−1,083	−3,000	−2,250	−750	−7,989	−7,813
2	−150	−756	−1,083	−3,000	−2,250	−750	−7,989	−7,642
3	−150	−756	−1,083	−3,000	−2,250	−750	−7,989	−7,473
4	−150	−756	−1,083	−3,000	−2,250	−750	−7,989	−7,309
5	−150	−756	−1,083	−3,150	−2,318	−765	−8,222	−7,356
6	−150	−756	−1,083	−3,150	−2,318	−765	−8,222	−7,194
7	−150	−756	−1,083	−3,150	−2,318	−765	−8,222	−7,036
8	−150	−756	−1,083	−3,150	−2,318	−765	−8,222	−6,881
9	−150	−756	−1,083	−3,308	−2,387	−780	−8,464	−6,928
10	−150	−756	−1,083	−3,308	−2,387	−780	−8,464	−6,776
11	−150	−756	−1,083	−3,308	−2,387	−780	−8,464	−6,626
12	−150	−756	−1,083	−3,308	−2,387	−780	−8,464	−6,481
Σ =	−1,800	−9,074	−12,997	−37,830	−27,818	−9,181	−98,700	−85,515

FV_{SF} ($) 10,000

$i_{TL} = i_{TL} P_{TL}$

$SF = PMT(i_{SF}, n, \nearrow P_{TL})$

$CR = PMT(i_{CR}, n, \nearrow P_{CR})$

$NPV(i_{IRR}, NCR_t = 1:NCF_t = n) = -\$85,515 \qquad = \Sigma(NCF_t)$

$L_t = L_t - 1(1 + g_L)t - 1$

$M_t = M_t - 1(1 + g_M)t - 1$

$OH_t = OH_t - 1(1 + g_{OH})t - 1$

$= -PV(i_{IRR}, n, \nearrow NCF_t)$

$= \Sigma(PV_t) = NPV$

$A_{equiv} = -\$8,211 = -PMT(i_{IRR}, n, NPV)$

TABLE 4.6

Example 3: Net Present Value

	Net Cash Flows (\$)		
t	P_1	P_2	P_3
0	0	0	0
1	−250	−1,000	−125
2	−250	−125	−125
3	−250	−125	−125
4	−250	−125	−125
5	−250	−125	−125
6	−250	−125	−125
7	−250	−125	−1,000
Σ	−1,750	−1,750	−1,750
NPV	−1,396	−1,523	−1,280
A_{equv}	250	273	229

Note: In all cases $r = 6\%$, $N = 7$,
$m = 1$, $i = 0.06$, and $n = 7$.

engineering economic analysis summarizes the net cash flow for the three alternatives and notes that the useful life of each alternative is unique and different from the others. The sum of the net cash flows reveals that Project 4 has the lowest total cost, Project 5 has the next lowest total cost, and Project 6 has the highest total cost. The net present value of the net cash flows for the three projects ranks the projects in the same order as the total net cash flows: Project 4 is the lowest and Project 6 is the highest. The net present value cannot be used to evaluate the economic preference for the projects because of the difference in each project's duration. The equivalent uniform recurring amount of the net present value for each project is calculated and tabulated in Table 4.7. Project 6 is found to be the low-cost alternative, although it had the highest total cost and the highest net present value.

Application to Reliability Based Life-Cycle Economic Analysis

Reliability analysis investigates multiple alternative design and sustainment solutions ranging from part selection options to make-buy decisions in system design and from capital acquisitions to maintenance practices options for critical parts.

Reliability analysis characterizes the occurrence of failure, mean time between failure, and its lower confidence limit, the demand function for

TABLE 4.7

Example 4: Equivalent
Uniform Recurring Amount

	Net Cash Flows ($)		
t	P_4	P_5	P_3
0	0	0	0
1	−250	−1,000	−125
2	−250	−125	−125
3	−250	−125	125
4	−250		−125
5	−250		−125
6			−125
7			−1,000
Σ	−1,250	−1,250	−1,750
NPV	−1,053	−1,160	−1,280
A_{equv}	250	434	229

Note: In all cases $r = 6\%$, $N = 7$,
$m = 1$, $i = 0.06$, and $n = 7$.

maintenance events, mean time to repair, and its upper confidence limit, and the duration and costs for maintenance and logistical events.

Engineering economic analyses fit cost objective functions that are combined with reliability failure, repair and logistics downtime math models to investigate the optimum least cost solution for an organization.

5

Reliability-Based Logistical Economic Analysis

Reliability-based logistical economic analysis is a subset of integrated logistical support analysis. Integrated logistical support analysis investigates the entire supply chain for a part from the manufacturer through all transportation and distribution points to the end user. Reliability-based logistical economic analysis applies the reliability math models to logistical events that are directly associated with part failure. Part level logistical economic analysis investigates the resources associated with part repair and system logistics downtime. The resources include labor skills mix, labor allocation to a task, maintenance planning, spare parts inventory, access to facility bays, and specialty tools, to name a few. Recall that all sustainability costs are caused by part failure. Part logistical economic analysis characterizes the sustainability costs. Remember costs occur at points in time: present amounts, uniform recurring amounts, and future amounts. Reliability analysis defines when sustainability costs will be incurred, mean time between failures (MTBF), the magnitude of the costs, and the duration of the cost events. The integration of reliability analysis with part logistical economic analysis is illustrated in Figure 5.1.

The failure math model, using the MTBF metric, defines the point in time that a part failure will occur. In engineering economics, the MTBF is equal to the number of compounding periods at the time that future amounts occur. The repair math model, using the mean-time-to-repair (MTTR) metric, defines the duration of the part repair event. The logistics math model, using the mean prerepair and postrepair logistics downtime metric, defines the duration of the logistics downtime.

Each of the reliability failure models provides the engineer with a deterministic estimate of the mean and its confidence level. The deterministic estimate of the mean is an adequate metric for evaluating the measure of the central tendency of all of the members of the population with uncertainty caused by the measure of dispersion. The confidence limit characterizes the expected magnitude of the reliability metric with a quantified level of uncertainty. Consider the MTBF: a cumulative failure math model evaluated for the MTBF is approximately 62%. That means that 62% of the parts will have failed by the time the population of parts has reached an operating time equal to the MTBF.

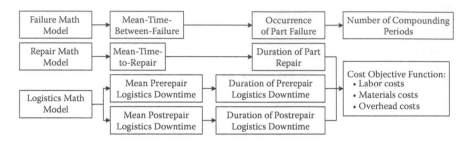

FIGURE 5.1
Integration of part reliability and logistical economic analysis.

Reliability analysis is interested in two calculations for confidence limits: the confidence limit of the sample and the confidence interval of the mean. The confidence limit of the sample is one sided. Reliability metrics fall in one of two categories: "higher is best" and "lower is best." Time to failure is higher is best; time to repair is lower is best. The lower confidence limit is applied to higher-is-best metrics, and the upper confidence limit is applied to lower-is-best metrics. The confidence interval of the mean is two sided. It has the limited application of performing a test of hypothesis for the statistical equivalence of two means at a specified confidence level. The null hypothesis is that there is no statistical difference between the two means. The alternate hypothesis is that the two means are statistically significantly different. We cannot distinguish between the two means regardless of their magnitudes if we accept the null hypothesis. We can rank order the two means if we reject the null hypothesis and accept the alternate hypothesis.

Failure Math Model

Fitting failure data to a reliability failure math model allows the characterization of the point estimate deterministic MTBF, the lower confidence limit of the MTBF, and the lower confidence limit of the time to failure. The expected time to the first part failure is the lower confidence limit of the time between failures.

For example, consider a part failure experiment that is fit to both the exponential and Weibull probability density functions for the part failure math models. The arithmetic mean time between failure is equal to 258 hours. The lower confidence limit for time to failure is the time in service that meets the condition that the cumulative part failure math model, $F(t)$, is equal to the level of significance, α; for example, the level of significance, α, is 0.05, given a confidence level equal to 95%. The exponential lower confidence limit for time to failure is equal to 13.25 hours, and the Weibull lower confidence limit

$\theta := 258.07$

$\eta := 168.94$

$\lambda := \dfrac{1}{\theta} = 3.875 \times 10^{-3}$

$\beta := 7.22$

$\gamma := 110$

$f_{exp}(t) := \lambda \cdot e^{-\lambda \cdot t}$

$$f_{3PW}(t) := \begin{vmatrix} 0 \text{ if } t < \gamma \\ \left(\dfrac{\beta}{\eta^{\beta}}\right) \cdot (t-\gamma)^{\beta-1} \cdot e^{-\left(\frac{t-\gamma}{\eta}\right)^{\beta}} \quad \text{if } t \geq \gamma \end{vmatrix}$$

$F_{exp}(t) := 1 - e^{-\lambda \cdot t}$

$$F_{3PW}(t) := \begin{vmatrix} 0 \text{ if } t < \gamma \\ 1 - e^{-\left(\frac{t-\gamma}{\eta}\right)^{\beta}} \quad \text{if } t \geq \gamma \end{vmatrix}$$

$LCL_{exp} := 13.25$

$LCL_{3PW} := 222$

$F_{exp}(LCL_{exp}) = 0.050$

$F_{3PW}(LCL_{3PW}) = 0.050$

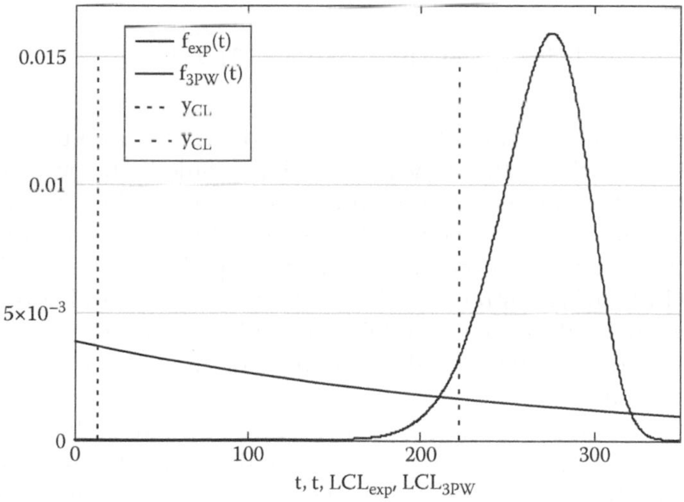

FIGURE 5.2
Comparative failure math models.

for time to failure is equal to 222 hours. The distinction between the lower confidence limits for time to failure is illustrated in the Mathcad worksheet in Figure 5.2.

The exponential failure math model and lower confidence limit for time to failure is represented by the dashed lines, and the Weibull failure math model and lower confidence limit for the time to failures is represented by the solid lines.

Why, you might ask, is the lower confidence limit for the time to failure not calculated using the following equation rather than finding the lower confidence limit value for which the cumulative failure math model is equal to alpha?

$$LCL = \mu + t_{\alpha,v}\sigma \tag{5.1}$$

The reason is that the cited equation only applies when the assumption that the data are normally distributed is accepted. Consider the exponential probability density function. The standard deviation is equal to the mean by definition. The lower confidence limit (LCL) will be negative for any value of the sampling statistic, Student's t. The equation for the Weibull lower confidence limit is expressed as

$$LCL = \gamma + \eta\sqrt[\beta]{-\ln(1-\alpha)} \tag{5.2}$$

also expressed as

$$LCL = \gamma + \eta\left(-\ln(1-\alpha)\right)^{1/\beta} = 221.962 \tag{5.3}$$

The reliability failure math model also fits a logistical metric known as the B10 life, the time at which 10% of the parts will have failed. The procedure for finding the B10 life is identical to the procedure to find the lower confidence limit for time to failure, where the B10 life is found to be that time for which the cumulative probability failure math model is equal to 0.1. The distinction between the B10 life is illustrated in the Mathcad worksheet in Figure 5.3.

The exponential failure math model and B10 life are represented by the dashed lines, and the Weibull failure math model and B10 life are represented by the solid lines.

The number of compounding periods can be expressed as the MTBF, the lower confidence limit of the time to failure, the B10 life, or the range of time to failure, minimum to maximum. In the next chapter, the application of the MTBF to determine the number of compounding periods will be presented for these methods.

The reliability failure math model is an essential tool to analytically determine a scheduled maintenance interval for the implementation of preventive maintenance. The fundamental concept of preventive maintenance is to remove and replace a part on a scheduled basis before the part is allowed to fail during scheduled operations of the system. On this point there is little disagreement throughout engineering and management. The problem is the determination of when the scheduled maintenance should occur in the life-cycle of the part and the system. If a part is replaced too frequently, then the

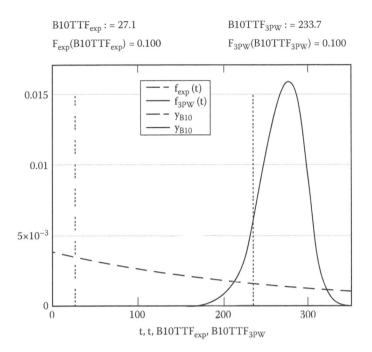

$B10TTF_{exp} := 27.1$

$F_{exp}(B10TTF_{exp}) = 0.100$

$B10TTF_{3PW} := 233.7$

$F_{3PW}(B10TTF_{3PW}) = 0.100$

FIGURE 5.3
Comparative failure math models with B10 life.

part fails to realize its full economic utility and excessive system downtimes occur. The solution to this dilemma is to identify an application of an organization's risk threshold, r, on the hazard function, $h(t)$, for the part. The risk threshold is defined as the risk of part failure that an organization is willing to accept for the deployment of the system on its next mission. Recall that the hazard function is the instantaneous failure rate for the part as it ages over time. The hazard function increases the longer the part is in service. The risk threshold can be expressed in terms of failures in time, i.e., failures per hour, failures per cycle, etc. The risk threshold applies to the system rather than the part for systems that are exclusively placed in a serial design configuration. The risk threshold applies to the part within parallel redundant design configurations.

For example, we fit the hazard function for the time-to-failure data from the previous example to the Weibull hazard function. The risk threshold is determined by management to be one part failure per month. The risk threshold is characterized as mission duration in hours, τ; number of missions per day, m; and the failure free duration, t_{ff}. Given that the mission duration is three hours, the number of missions per day is 2, and the failure free duration is 20 operating days in a month. The risk threshold is calculated as the inverse of the product of the mission duration, number of missions per

$$\theta := 258.07 \qquad\qquad \eta := 168.94 \qquad \tau := 3$$

$$\lambda := \frac{1}{\theta} = 3.875 \times 10^{-3} \qquad\qquad \beta := 7.22 \qquad m := 2$$

$$\gamma := 110 \qquad t_{ff} := 20$$

$$r := \frac{1}{\tau \cdot m \cdot t_{ff}} = 8.333 \times 10^{-3}$$

$$h(t) := \begin{vmatrix} 0 \text{ if } t < \gamma \\[2mm] \left(\dfrac{\beta}{\eta^{\beta}}\right) \cdot (t - \gamma)^{\beta-1} \text{ if } t \ge \gamma \end{vmatrix} \qquad\qquad F(t) := \begin{vmatrix} 0 \text{ if } t < \gamma \\[2mm] -\left(\dfrac{t-\gamma}{\eta}\right)^{\beta} \\ 1 - e \qquad \text{ if } t \ge \gamma \end{vmatrix}$$

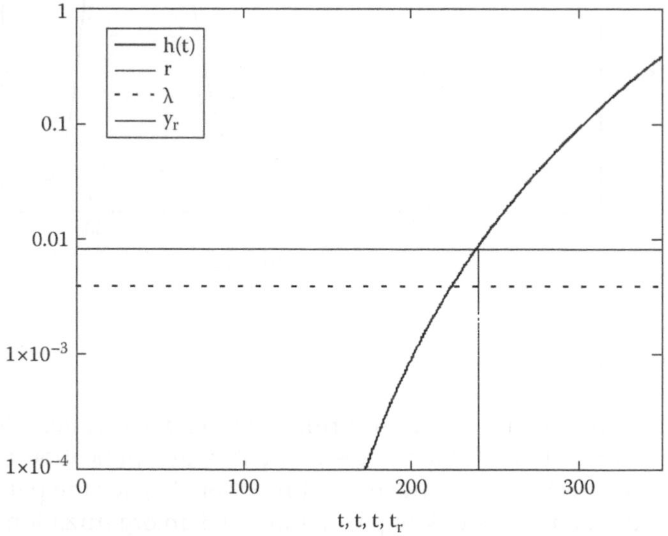

FIGURE 5.4
Comparative hazard functions with risk threshold.

day, and the failure free duration, $1/120 = 0.00833$ failures per hour. The risk-based scheduled maintenance interval, t_r, is the time that meets the condition that the hazard function is equal to the risk threshold, $h(t_r) = r$. Finding the value of the scheduled maintenance interval can be performed in Excel using goal seek and in Mathcad using the given routine that sets the risk threshold to the Boolean equivalent of the hazard function followed by the Find(t) evaluation. The Mathcad approach is illustrated in Figure 5.4.

Given

$$r = \left(\frac{\beta}{\eta^{\beta}}\right)(t - \gamma)^{\beta-1} \qquad\qquad (5.4)$$

Boolean equivalence

$$\text{Find}(t) \rightarrow 239.893326\ldots$$

$$t_r = 239.89$$

$$h(t_r) = 8.332 \times 10^{-3}$$

The Weibull hazard function is the bold increasing line, the risk threshold is the horizontal solid line, the risk-based scheduled maintenance interval is the vertical solid line, and the horizontal dashed line is the exponential distribution failure rate. Observe that the exponential failure rate is constant and can either be above or below the risk threshold. As such it is not capable of determining a risk-based scheduled maintenance interval.

The risk-based scheduled maintenance interval calculates the time in service at which the part will exceed the risk threshold on the next mission. As a planning tool, the organization must determine the advanced time in which the systems maintenance action is coordinated with operations and all of the maintenance resources are acquired and ready for use. The maintenance cycle, logistics downtime, and repair time can begin at or before the risk-based scheduled maintenance interval.

Repair Math Model

The repair math model provides the MTTR and the upper confidence limit for the time to repair. Two approaches have been provided for the repair math model: the log-normal probability distribution and the Weibull probability distribution. Absent empirical data, the triangular distribution can be used as a placeholder distribution. Recall that the resources to perform repair experiments are finite and limited to the highest priority critical parts. It is not uncommon that resources will be applied to fit only the failure math model.

The duration of part time to repair is expressed as the MTTR, the upper confidence limit for the time to repair, and the range between the minimum and maximum time to repair. The importance of the duration of the time to repair is that it is the multiplier used to calculate maintenance cost future amounts, to include labor, specialty tools, and overhead. The future amount for maintenance costs is equal to the product of the metric for time to repair and the hourly labor rate. For example, given an MTTR of four hours, a labor rate of $50 per hour, a utilization rate of $10 per hour for an overhead crane,

TABLE 5.1

Time-to-Repair Cost Analysis

	MTTR (hr)	Cost ($)
Labor rate (at $50/hr)	4	200
Overhead crane (at $10/hr)	4	40
Facility bay (at $15/hr)	4	60
Future amount ($/repair event)		300

and a utilization rate of $15 per hour for a facility bay, the future amount for the maintenance action is shown in Table 5.1.

Mean Maintenance Time

The calculation for MTTR and its upper confidence limit is defined to be the baseline reliability function. As noted before, mean maintenance time can include maintenance actions performed in a facility and maintenance actions performed at the location of the system. The organization can develop a critical part maintenance list that identifies part repair actions that must be better understood than what the baseline MTTR provides. For such parts, the mean maintenance time should be calculated for both maintenance scenarios. Otherwise, the MTTR will be the mean of the means of two populations, and the measure of dispersion will not reflect the measure of dispersion for each population. This will be a weighted average MTTR that calculates the mean time to repair for maintenance actions performed in the facility and the maintenance actions performed at the system site using the frequency of each as the weighting factors.

Corrective and Preventive Maintenance

As a preventive maintenance program is implemented, the MTTR will be comprised of increasing preventive maintenance actions and decreasing corrective maintenance actions. The characterization of the corrective maintenance actions will continue to be comprised of the two environment scenarios. Preventive maintenance will not realistically eliminate all corrective maintenance requirements. This can be empirically demonstrated by calculating the cumulative maintenance repair model for the risk-based scheduled maintenance interval. Using the example data above, the cumulative maintenance repair model evaluated at 239.89 hours is equal to 0.134, or 14%. In other words, the implementation of preventive maintenance reduced the incidence of corrective maintenance actions from 100% to 14%. The preventive maintenance MTTR will differ little from the MTTR for corrective maintenance actions performed in the maintenance facility.

Reduction of the corrective maintenance actions performed at the system location will see the baseline MTTR decrease by as much as 25%. This provides an interesting bonus to the organization. Consider the example in which the MTTR is equal to four hours. A 25% reduction in the MTTR, to three hours, is equivalent to adding one hour of labor for the repair maintenance team. This equals an increase in equivalent man-hours available to the organization without an increase in personnel and their costs to the organization.

Logistics Downtime Math Model

The logistics downtime math models provide two multipliers for the future amount of the maintenance action: the mean prerepair logistics downtime × the hourly rate for labor and overhead, and the mean postrepair logistics downtime × the hourly rate for labor. Consider the following example: the prerepair logistics downtime is six hours, and the postrepair logistics downtime is one hour. The hourly labor rate for prerepair logistics downtime is $50, and $40 for postrepair logistics downtime. In addition to labor, prerepair logistics downtime requires a retrieval vehicle that has an hourly rate of $110 an hour. The future amount for logistics downtime is shown in Table 5.2.

Impact of Preventive Maintenance on Logistics Downtime

Prerepair logistics downtime is comprised of tasks that are associated with unplanned demand for maintenance. We have observed that these tasks include waiting time for labor, facilities, and specialty tools, in addition to the time required to transport the system to the maintenance facility or to transport the maintenance capability to the system. Preventive maintenance is equivalent to making an appointment. The time delay associated

TABLE 5.2

Logistic Downtime Cost Analysis

	Cost ($)
Prerepair logistics downtime = 6 hr	
Labor (at $50/hr)	300
Retrieval vehicle (at $110/hr)	660
Future amount (per repair event)	960
Postrepair logistics downtime = 1 hr	
Labor (at $40/hr)	40
Future amount (per repair event)	40

with transporting the system to the maintenance facility or the maintenance capability to the system is eliminated. The waiting time for labor, facilities, and specialty tools is either eliminated or minimized. A few prerepair logistics downtime events will remain unchanged or reduced, to include failure reporting, maintenance documentation, etc. Anecdotal experience has demonstrated that the reduction in prerepair logistics downtime has always exceeded 50% and has reached as much as 90%.

Postrepair logistics downtime may be reduced but to a smaller degree than for prerepair logistics downtime.

Special Cause Variability in Logistics Downtime

The logistics downtime math models that have been presented so far address the common cause variability for logistics downtime events. For example, the retrieval of the system to the maintenance facility has measures of central tendency and dispersion. The measure of dispersion describes the common cause variability of the event. Common cause variability is intrinsic to the event. Special cause variability is comprised of extrinsic factors. Consider a mining haul truck that has a flat tire in the mine. The prerepair logistics events associated with transporting a spare tire and tire change equipment can be measured with a mean and a standard deviation. The standard deviation characterizes the common cause variability of the event. Consider now that the haul truck not only has a flat tire, but swerved and tipped over on its side. The prerepair logistics events associated with acquiring a crane to lift the truck back up on its wheel base is not intrinsic to the prerepair logistics events.

Special cause variability in field sustainability has typically been categorized as follows:

1. Lack of a spare part: Prerepair logistics downtime is increased by the time required to order and receive a spare part, which may take days, while the system remains in a down state until the spare part arrives.
2. Lack of a labor skill: Prerepair logistics downtime is increased by the inability of the maintenance organization to perform the repair events, requiring a search for outside labor skill, which may be hampered by limited availability, etc.
3. Lack of a specialty tool: Prerepair logistics downtime is increased by the limited availability for a specialty tool, for example, an overhead crane that is currently in use by another system, computer-based system diagnostic equipment that is in a down state, etc.

4. Lack of an efficient management information system: Prerepair logistics downtime is increased by the maintenance organization's inefficient approach to coordinating resources, to include excessive delays in finding and acquiring a spare part, organizing the maintenance repair team, etc.

5. Lack of sufficient maintenance facilities: Prerepair logistics downtime is increased by the waiting time required to have access to the maintenance facility; the maintenance facility is inadequate due to technological innovation.

6. Lack of a cooperative relationship between maintenance and operations management: Prerepair logistics downtime is increased by the inability of maintenance and operations managers to commit to maintenance planning, maintenance funding, etc.

A preventive maintenance program that is developed and implemented by maintenance and operations management jointly achieves elimination of special causes in a variety of logistics downtime scenarios.

Cost Objective Function

Reliability failure math models provide the multipliers for cost calculations for maintenance and logistics downtime events. The engineer and manager must understand what constitutes a cost. There are four cost categories: labor, materials, overhead, and lost opportunity. Labor, materials, and overhead costs are typically well understood by an organization and are measured as expenses that are captured by the management information system for accounting and budgeting. Lost opportunity cost is not well understood because it is not included in the management information system. Yet it is the most significant cost to the organization.

Labor Costs

There are two labor costs to an organization: the cost of an employee to the organization and the cost of a labor hour or an employee category. An employee receives direct compensation in the form of a wage or salary. The organization incurs compensation costs referred to as variable fringe benefits, to include: workman's compensation, Social Security and Medicare contributions, medical coverage, tuition reimbursement, etc. The variable fringe benefit rate is calculated as a percentage of the employee's base wage or salary that is revised periodically. Fringe benefits can also be fixed. For example, the compensation cost is a constant amount that is added to the employee's base wage or salary, to include a company vehicle, membership in a health club. Total compensation costs with fringe benefits, $L_{w/FB}$, are calculated as the sum of the employee's base wage or salary, the product of the variable fringe benefit rate, f_v, and the base wage or salary, I, plus the fixed fringe benefit, as expressed in the following equation.

$$L_{w/FB} = I + f_v(I) + f_f = I(1 + f_v) + f_f \tag{5.5}$$

The total labor cost of an employee to the organization includes the direct overhead costs for supporting infrastructure. Direct overhead (DOH) is comprised of labor, materials, and overhead expenses that support the employee's working group. Consider a maintenance department that incurs the following costs:

- Labor: maintenance manager, administrative assistants, safety engineer, training staff, and janitorial employees
- Materials: office supplies, desks and furniture, office equipment, lockers, and shower rooms
- Overhead: utilities, allocation of facility costs to offices, and employee rooms

These costs are incurred by the organization for having a maintenance organization and are allocated to the hourly cost of an employee to the organization in the form of a variable direct overhead rate.

Total labor costs with direct overhead, $L_{w/DOH}$, are calculated as the sum of the labor with fringe benefits (FB) and the product of the variable direct overhead factor, F_{DOH}, as expressed in the following equation.

$$L_{w/DOH} = L_{w/FB} + f_{DOH}(L_{w/FB}) = L_{w/FB}(1 + f_{DOH}) \tag{5.6}$$

The total labor cost of an employee to the organization also includes the indirect overhead (IOH) costs for the organization's infrastructure above the employees' operating group. Indirect overhead is comprised of labor, materials, and overhead expenses that support the general management of the organization. Consider an organization that incurs the following costs:

- Labor: management, human resources employees, finance and accounting employees, marketing and sale employees, administrative employees, staff employees
- Materials: office supplies, desks and furniture, office equipment, executive dining rooms, sales materials, business forms, business travel, consultants
- Overhead: utilities, offices, corporate jets

TABLE 5.3

Cost of an Employee to the Organization

	Fully Burdened Direct Labor Rate ($/hr)			
Wage ($/hr)	Var fringe 25% ($)	Dir O/H 137% ($)	Indir O/H 18% ($)	Burden Rate
10.00	12.50[a]	29.63[b]	34.96[c]	3.496[d]

[a] $12.50 = $10.00 × (1 + 25%).
[b] $29.63 = $12.50 × (1 + 1.37).
[c] $34.96 = $29.63 × (1 + 1.18).
[d] 3.496 = $34.96/$10.00.

Total labor costs with indirect direct overhead, L, are calculated as the sum of the labor cost with direct overhead, L_{DOH}; and the product of the variable indirect overhead factor, f_{DOH}; and the labor cost with direct overhead, L_{DOH}, as expressed in the following equation.

$$L = L_{w/DOH} + f_{IOH}\left(L_{w/DOH}\right) = L_{w/DOH}\left(1 + f_{IOH}\right) \qquad (5.7)$$

Most organizations compute a labor burden rate factor, f_{BR}, as illustrated in Table 5.3. The labor burden rate factor simplifies the cost of employee labor to the organization to the product of the labor burden rate factor and the wage or salary, I, as expressed in the following equation.

$$L = f_{BR}\left(I\right) \qquad (5.8)$$

The cost of an employee to the organization is illustrated in Table 5.3.

The cost of a labor skill category to the organization addresses the fact that employees are paid for time not applied to their job. Employees are provided with annual leave, holidays, and sick days. Employees are also assigned to administrative time away from their work site, to include safety training, job training, etc. Labor law defines a standard 40-hour work week for hourly employees. Time worked in excess of 40 hours in a week is paid a 50% premium, called "time and a half." Holiday work is also paid a premium that can be as much as 100%, called "two-times wage." Salaried employees are paid a fixed amount without regard to hours worked. A standard work year is 2,080 hours (52 weeks times 40 hours). Leave and holidays vary by employee grade, apprentice, master technician, and years of longevity, typically increments of five years employment with the organization. The following table describes the work and nonwork times for a straight work week, 2,080 hours/year, two weeks leave, 80 hours/year, 10 paid holidays, 80 hours/year, three paid sick leave days, 24 hours/year, and 144 administrative hours for various reasons in the year. Table 5.4 calculates that the employee works 1,792 hours per year while being paid for 2,080 hours. The ration of work hours

TABLE 5.4

Billable Hours Utilization Factor

	Hours
Year	2,080
Leave	80
Holiday	80
Sick	24
Admin	144
Direct	1,752
Factor (%)	84.23
Emp Equiv	1.19

to total hours is the employee's utilization factor, f_u = 86.15% in the example. The organization schedules for one full-time employee category, mechanic, maintenance planner, spare parts clerk, for the year. The organization's employee utilization factor calculates that 1.2 employees in a required category are required to perform 2,080 hours per year, $1/f_u$ = 1/0.8615 = 1.2.

Organizations with large numbers of employees distributed across many labor categories for several labor grades simplify labor cost calculations with standard costs. Standard costs are weighted averages of the hourly labor rate paid to the employee; for example all master electricians are estimated at the same hourly rate with all factors applied. The standard labor costs are revised periodically, as are the burden factors.

Multiple Shifts

The total labor cost to the organization that schedules multiple shifts includes a shift premium. Second shift (swing or night shift) and third shift (graveyard shift) have successively higher shift premiums. The calculation of the total labor cost to the organization for an employee is illustrated in Table 5.5.

TABLE 5.5

Multishift Fully Burdened Direct Labor Rate

Wage ($/hr)	Shift Premium (%)	Var Fringe 25% ($)	Dir O/H 137% ($)	Indir O/H 18% ($)	Burden Rate
10.00	100	12.50[b]	29.63[c]	34.96[d]	3.496[e]
11.50[a]	115	14.38	34.07	40.20	3.496
12.50	125	15.63	37.03	43.70	3.496

[a] $11.50 = $10.00 × 115%.
[b] $12.50 = $10.00 × (1 + 25%).
[c] $29.63 = $12.50 × (1 + 1.37).
[d] $34.96 = $29.63 × (1 + 1.18).
[e] 3.496 = $34.96/$10.00.

The second and third shift base hourly wage is the product of the straight time base hourly wage and the shift premium. Then the burden factors are applied for direct and indirect overhead. The burden rate for the three shifts remains constant. The cost of the labor category to the organization is calculated by the utilization factor just as with the straight time employee.

Labor Cost per Hour for Multishift Operations

All shifts are not equal. Nor can maintenance logistical resources be assumed to occur on a specific shift. Often maintenance tasks extend across shifts. The labor cost for an employee category is the expected value of the labor costs per employee category for the shifts. The shift labor distribution is the number of the employees of a labor category assigned to a shift divided by the number of the employees of a labor category. Assume that 50% of all supervisors are assigned to the first shift, 33% are assigned to the second shift, and 17% are assigned to the third shift, as shown in the following supervisor shift distribution, $f(x) \in \{0.500, 0.333, 0.167\}$.

The annualized base standard cost for the income for an employee category is used to calculate the base hourly rate paid to the employee and the base hourly rate for the employee category. For example, the supervisor labor category has a base standard cost of $60,000/year on first shift, or an hourly rate of $28.85 ($60,000/2,080 = $28.85). The base standard hourly cost for the supervisor labor category is the hourly rate divided by the utilization factor, $28.85/0.8615 = $33.48. The application of the burden rate to the supervisor labor category is the hourly cost to the organization, 3.50 × $33.48 = $117.05. The application of shift premiums to the base hourly labor rates calculates $134.60 and $146.31, respectively. The expected value for the supervisor labor category is equal to the sum of the products of the supervisor shift distribution and the base hourly labor rates, $127.08, as shown in Tables 5.6 and 5.7.

The computations for the supervisor labor category for the three shifts are shown in Table 5.8.

The future amount for a supervisor for a repair and logistics downtime event is the product of the reliability multiplier (MTTR, prerepair logistic downtime, and postrepair logistic downtime), measured in hours, and the supervisor labor category hourly rate.

TABLE 5.6

Labor Shift Distribution

Shift	$f(x)$
1	0.500
2	0.333
3	0.167

TABLE 5.7

Expected Value for the Supervisor
Labor Category

$\mu L = \Sigma x f(x)$	
X ($)	xf(x) ($)
119.72	59.86
137.68	45.85
149.65	24.99
μL	130.70

TABLE 5.8

Fully Burdened Hourly Labor Rate for Supervisor Labor Category

Labor		Wage/hr ($)		Shift Premium (%)	Burden Rate ($/hr)	Labor Expense ($)		Annual STD
Category	Shift	Emp	Org			Shift	Category	Pay ($)
Supervisor	1	28.85	34.25	100	3.50	119.72	130.70	60,000
	2	33.17	39.38	115		137.68		
	3	36.06	42.81	125		149.65		

Materials Costs

A materials cost to the organization is comprised of the vendor's price, the raw cost, and a materials overhead factor. The materials overhead costs include

- Labor: shipping and receiving labor expenses, inventory clerk labor expenses
- Materials: office supplies, desks and furniture, storage structure, packing and kitting containers
- Overhead: utilities, allocation of facility costs to offices, and storage rooms

Consider a part that has a vendor price, $P = \$10.00$. The overhead rate, f_{OH}, is 46%, and the fully burdened cost of the part to the organization is the vendor price plus the product of the overhead rate and the vendor price, as expressed in the following equation.

$$M = P + f_{OH}(P) = P(1 + f_{OH}) \tag{5.9}$$

The materials burden rate is equal to the fully burdened cost of the part divided by the price. The calculations for the fully burdened materials cost and burden rate are provided in Table 5.9.

TABLE 5.9

Fully Burdened Materials Cost

Raw Cost ($)	O/H 46% ($)	Burden Rate
10.00	14.60	1.460

Spare Part Cost to the Organization

The cost of a spare part and the occurrence of the spare part acquisition to the organization are critical logistics economic factors.

Spare Parts Acquisition Strategies

There are several approaches to spare parts acquisition. The most common is the random unit order. The other strategies include: economic order quantities (EOQ), just in time (JIT), and emergency spares. Each differs in the unit price and turnaround time (duration between the order and the delivery).

Random Unit Orders

The random unit order strategy acquires a spare part when the spare in inventory is used. The vendor price is the retail price. The turnaround time and its relationship to the occurrence of the demand for the spare part upon the next part failure is determined by the MTBF and the lower confidence limit of the time to failure. Consider a part with an MTBF equal to 1,000 hours and the lower confidence limit of the time to failure is 900 hours. The turnaround time for the part is two weeks. The MTBF must be expressed in the same units of time for the turnaround time. For example, a haul truck is operated on three shifts a day, seven days a week. A 30-minute service time occurs between each shift, yielding 22.5 possible operating hours per day or 157.5 hours per week. The demand rate for the part using the lower confidence limit for the time to failure is 1 per 5.7 weeks at 100% operational availability (an unrealistic level). The inventory carrying cost for the part begins upon receipt at the end of the second week and ends 3.7 weeks later. Inventory carrying cost is comparable to overhead in composition. Inventory carrying costs are the expenses incurred by the organization to receive, inspect, and store a part until it is installed on a system. The cost of operating capital is tied up in the acquisition cost of the part. Assume that two parts are installed on each system and there are 20 systems. The random unit spare part strategy will create 40 spare part transactions corresponding to 40 part failures. The random unit spare parts strategy is in equilibrium when the spare part demand approaches the deterministic metrics for time to failure.

The random unit spare parts strategy approaches equilibrium using the first-in-first-out approach, assuming that the measure of dispersion for time to failure is accurately characterized and is relatively small.

Economic Order Quantity

The economic order quantity spare parts strategy applies the understanding of the part demand rate with the understanding of the part acquisition cycle to calculate the cost optimum spare part quantity. The objective of the economic order quantity spare parts strategy is to achieve the minimum inventory cost at an acceptable risk. The minimum inventory cost determines the optimum spare parts quantity that minimizes the unit cost of the parts through quantity discounts, minimizes inventory carrying costs, and prevents the supply of required spare parts from depleting. EOQ strategy has existed for a century. It improves on the random unit spare parts strategy by consolidating the acquisition of spare parts rather than treating each acquisition as an individual event. Variations of the EOQ strategy are plentiful, but the objective remains the same. The time to failure metric is the essential factor in determining the optimal order quantity for parts. That fact is consistent across all variations of the economic order quantity strategy. The supply chain metrics are optimized to achieve the spare part demand function defined by the time to failure metric and provide the substance for variations of the economic order quantity strategy.

Just-in-Time Spare Parts Strategy for Preventive Maintenance

The just-in-time spare parts strategy is a refinement of the random unit order spare parts strategy and is an element of a preventive maintenance practice. It seeks to achieve the ideal situation where the organization knows the exact date that the spare part is required and is able to manage the supply chain to assure that the part arrives on that date. The acquisition cost of the spare part to the organization is limited to the cost-burdened vendor price. The risk-based time to scheduled maintenance derived from the hazard function for time to failure is the metric that enables an organization to implement a JIT spare parts strategy.

As mentioned earlier, preventive maintenance practices do not eliminate all unscheduled corrective maintenance actions. The JIT spare parts strategy must be tempered by a conventional random unit order spare parts strategy to minimize the risk to the organization that a part will not be available when required.

Emergency Spare Part

Emergency spare parts are not a strategy but rather the consequences of an inefficient spare parts strategy. The emergency spare part is required when a spare part is not available for a critical system. Crashing the supply chain process to acquire a part as soon as possible in order to minimize the downtime of a system carries a very high premium cost. Consider the mining example where we have a fleet of mining haul trucks. The haul trucks are loaded in the mine by a single excavator, a dragline, shovel, or front end loader. The impact of a single haul truck in a down state is less severe than the excavator being in a down state. The loss of a single haul truck reduces the mine's productivity; the loss of the excavator ceases all mine productivity. Assume that the organization does not have a critical spare part, and assume that the turnaround time for that spare part is several weeks. The mine cannot afford the lost opportunity costs that would be incurred for the turnaround time. The organization assigns a team to (1) negotiate an expedited acquisition from their supplier, (2) negotiate an acquisition from a competitor or another division within the organization, or (3) design and manufacture the part internally or using a contract supplier. In the first case, the vendor may be able to provide an emergency spare by ramping up the acquisition process and passing on the increased costs to the mine. In the second case, the competitor may cooperate, which will impose a quid pro quo, and the other division within the organization may call on upper management to require the transfer of the part, leading to extended time delays. In the third case, the expedited design and manufacture will incur labor, materials, and overhead costs that will exceed the normal acquisition cost. In all three cases, the cost of the emergency spare team will incur labor, materials and overhead costs. The emergency spare part multiplier can range from 200% to 1,000%.

Specialty Tools

Specialty tools are categories of machines, equipment, and tools that have specific applications to a part's maintenance, including prerepair logistics events, repair events, and postrepair logistics events. The most common specialty tool is the mine maintenance truck. Examples of other specialty tools include systems for replacement of off-road vehicle tires, overhead cranes in a maintenance facility, computerized diagnostic machines, etc. Specialty tools are capital investments. The demand for specialty tools is determined by the maintainability analysis. Initially, the maintainability analysis was performed to identify the tasks needed to perform a part maintenance event defining the repair experiment in order to fit the part repair math model. The maintainability analysis also includes the determination for the requirement of specialty tools to perform a part maintenance event.

The logistics economic analysis for part maintenance requires one to determine the hourly rate of the specialty tool. The hourly cost to the organization for the use of a specialty tool requires one to apply engineering economics to determine the labor, materials, and overhead costs for the operation and maintenance of the specialty tool over its useful life. The approach to calculate the hourly cost is presented in the following example.

Example: Retrieval Vehicle

The hourly rate for a capital asset is equal to the annual cost incurred by the organization divided by the operating hours in each year of the useful life of the asset. The hourly rate may be calculated for time periods less than annually depending on the frequency of the application of cost escalators. The hourly rate at which any capital asset is charged against a particular maintenance event requires development of a cash flow timeline for capital costs and operating and maintenance costs, as presented in Table 5.10.

TABLE 5.10

Economic Information for Retrieval Vehicle

Capital Costs Estimation		Operating and Maintenance Costs Estimation	
P_{Vehicle} (acquisition cost, $)	−28,000	Exp_{DLO} (baseline direct labor operations, $/yr)	−411,032
			3.33
r_{Loan} (loan APR, %)	4.50	g_{DLO} (DLO cost escalation rate	2
N_{Loan} (yr)	3	APR, %)	0.016650
m_{Loan} (pp/yr)	4	m_{DLO} (DLO cost escalations/yr)	
n_{Loan} (pp)	12	i_{DLO} (effective DLO cost escalation	
i_{Loan} (i/pp)	0.011250	rate)	
PMT_{Loan} (iP, $)	−315		
		Exp_{DLM} (baseline maintenance labor	−57,088
r_{SF} (sinking fund APR, %)	8.25	operations, $/yr)	4.15
N_{SF} (yr)	3	g_{DLM} (cost escalation rate APR, %)	2
m_{SF} (pp/yr)	12	m_{DLM} (cost escalations/yr)	0.020750
n_{SF} (pp)	36	i_{DLM} (effective cost escalation rate)	
i_{SF} (i/pp)	0.006875		
PMT_{SF} ($)	−688	$\text{Exp}_{\text{Matls}}$ (baseline materials cost,	−13,330
		$/yr)	5.67
f_{SV} (salvage value factor)	0.8	g_{Matls} (cost escalation rate APR, %)	12
N_{Vehicle} (useful life)	5	m_{Matls} (cost escalations/yr)	0.004725
F_{Vehicle} (salvage value, $)	9,175	i_{Matls} (effective cost escalation rate)	
N_{Dep} (yr)	5	$\text{Exp}_{\text{DO/H}}$ (baseline direct overhead	−32,593
m_{Dep} (pp/yr)	12	cost, $/yr)	1.67
n_{Dep} (mon)	60	$g_{\text{DO/H}}$ (cost escalation rate APR, %)	1
PMT_{Dep} ($)	−314	$m_{\text{DO/H}}$ (cost escalations/yr)	0.016670
		$i_{\text{DO/H}}$ (effective cost escalation rate)	
r_{CT} (capital tax rate APR, %)	2.15		
m_{CT} (pp/yr)	12	r_{FTax} (federal tax rate, %)	39
i_{CT} (capital tax rate/pp)	0.001792	r_{STax} (state tax rate, %)	6.60

The capital cost estimation for the retrieval vehicle includes the following cost events:

- Term loan: to finance the acquisition cost of the retrieval vehicle
  - Present amount of the asset: $P_{\text{Vehicle}} = -\$28,000^*$ in the example
  - Terms and conditions of the term loan: 3-year term loan, N_{Loan}, at 4.5% APR, r_{Loan}, with interest paid quarterly, $m_{\text{Loan}} = 4$ (the number of payment periods over the term of the loan, $n_{\text{Loan}} = m_{\text{Loan}}N_{\text{Loan}} = 12$ quarters, and the interest rate payable quarterly, $i_{\text{Loan}} = r_{\text{Loan}}/m_{\text{Loan}} = 0.01125$)
  - Interest expense per payment period: $\text{PMT}_{\text{Loan}} = P_{\text{Loan}} \times i_{\text{Loan}} = -\$315/\text{quarter}$
- Sinking fund investment: to pay off the term loan at the end of the loan term
  - Terms and conditions of the sinking fund: three year duration, N_{SF}, at 8.25% APR, r_{SF}, with interest compounded monthly, $m_{\text{SF}} = 12$ (the interest rate compounded monthly, $i_{\text{SF}} = r_{\text{SF}}/m_{\text{SF}} = 0.006875$, number of compounding periods over the duration of the sinking fund, $n_{\text{SF}} = m_{\text{SF}}N_{\text{SF}} = 36$ months, and
  - Monthly investment, $\text{PMT}_{\text{SF}} = \text{PMT}(i_{\text{sf}}, n_{\text{sf}}, F_{\text{Loan}}) = \text{PMT}(0.006875, 36, -\$28,000) = -\$688/\text{month})$
- Salvage value of the asset: the residual value of the retrieval vehicle at the end of its useful life
  - Salvage value factor is the multiplier of the acquisition cost that estimates the residual value of the retrieval vehicle annually: $f_{\text{SV}} = 0.8$
  - Useful life of the retrieval vehicle: $N_{\text{Vehicle}} = 5$ years
  - Future amount of the retrieval vehicle:

$$F_{\text{Vehicle}} = P_{\text{Vehicle}}f_{\text{SV}}^{N_{\text{Vehicle}}} = (\$28,000)0.8^5 = \$9,175$$

at the end of year 5
- Depreciation of the asset: the non–cash flow expense that amortizes the retrieval vehicle to determine the book value at the end of each year of the useful life
  - Useful life of the retrieval vehicle: $N_{\text{Vehicle}} = 5$ years
  - Frequency of the depreciation expense: $m_{\text{Dep}} = 12$ (the number of depreciation periods, $n_{\text{Dep}} = m_{\text{Dep}}N_{\text{Dep}} = 60$)

---

* Sign convention for cash events: (–) designates uses of cash, expenses; (+) designates sources of cash, proceeds of a loan, tax credits.

- Depreciation balance: $V_{Dep} = P_{Vehicle} - F_{Vehicle} = \$28{,}000 - \$9{,}175 = \$18{,}825$
- Depreciation expense per period: $PMT_{Dep} = V_{Dep}/n_{Dep} = \$18{,}825/60 = -\$314$
- Capital tax paid on the book value of capital assets
  - Book value of a capital asset at any point in time is equal to the present amount valuation of the asset minus the cumulative depreciation expense taken against the asset: $V_{Book} = P_{Vehicle} - \Sigma PMT_{Dep}$
  - Frequency of the capital tax payments: $m_{CT} = 12$
  - Capital tax rate: $r_{CT} = 2.15\%$ APR (the capital tax rate per payment period, $i_{CT} = r_{CT}/m_{CT} = 0.001792$)
  - Capital tax expense: $PMT_{CT} = i_{CT}V_{Book}$

The baseline operating and maintenance cost (O&M) estimations for the retrieval vehicle include the following cost events:

- Annual estimated operating expenses for the retrieval vehicle
  - Sum of all direct labor expenses for the first year: $EXP_{DLO} = -\$401{,}857$ (the cost escalator for the direct labor for operations (DLO), $g_{DLO} = 3.33\%$ APR, is applied semiannually, $m_{DLO} = 2$, for an effective direct labor for operations cost escalation rate, $i_{DLO} = 0.01665$)
- Annual estimated maintenance expenses for the retrieval vehicle
  - Sum of all direct labor expenses for the first year: $EXP_{DLM} = -\$55{,}814$ (the cost escalator for the direct labor for maintenance (DLM), $g_{DLM} = 4.15\%$ APR, is applied semiannually, $m_{DLM} = 2$, for an effective direct labor for maintenance cost escalation rate, $i_{DLM} = 0.02075$)
  - Sum of all materials expenses for the first year: $EXP_{Matls} = -\$13{,}330$ (the cost escalator for the materials, $g_{Matls} = 5.67\%$ APR, is applied monthly, $m_{Matls} = 12$, for an effective materials cost escalation rate, $i_{Matls} = 0.004725$)
- Annual estimated direct overhead expenses for the retrieval vehicle
  - Sum of all direct overhead expenses for the first year: $EXP_{DLM} = -\$32{,}593$ (the cost escalator for the direct overhead for maintenance, $g_{DLM} = 1.67\%$ APR, is applied annually, $m_{DLM} = 1$, for an effective direct overhead cost escalation rate, $i_{DLM} = 0.0167$)
- Tax credits for tax-deductible expenses for the retrieval vehicle (Table 5.10)

- Federal tax credit: federal tax rate, $r_{FTax} = 39\%$, applied to the sum of tax-deductible expenses for capital, operating, and maintenance costs
- State tax credit: state tax rate, $r_{STax} = 6.6\%$, applied to the sum of tax-deductible expenses for capital, operating, and maintenance costs adjusted for the federal tax credit

Cash Flow Timeline

The cash flow timeline, presented in Table 5.11, records the estimation of capital, operating, and maintenance costs for the retrieval vehicle over its five-year useful life. The analyst chose monthly time units for the cash flow timeline because the cost escalation factor for direct materials is applied monthly. The cash flow timeline includes the distinction between tax-deductible expenses (non-bold columns) and non–tax-deductible expenses (bold columns).

Capital Costs

- Column 1 specifies the time units (in months).
- The loan and capital acquisition shown in the table title net out for a net cash flow equal to zero. Capital sources and uses of cash are not tax-deductible expenses and are actual cash transactions.
- Column 2 records the estimations and quarterly occurrences for the interest payments for the term loan through the end of the third year, month 36 (PMT_{Loan}). Interest payments on loans are tax-deductible expenses and actual cash transactions.
- The salvage value of the retrieval vehicle at the end of its useful life, the end of the fifth year, month 60 ($F_{Vehicle}$), is given in the table title. Salvage value is a taxable credit and not an actual cash transaction.
- Column 3 records the monthly investment in the sinking fund through the end of the third year, month 36 (PMT_{SF}). Payments to a sinking fund are not tax-deductible expenses; however, they are actual cash transactions.
- Column 4 is an internal working table that records the monthly investment balance in the sinking fund (Bal_{SF}) and shows that the investment interest compounded monthly provides sufficient funds to pay the term loan at the end of the third year. The monthly investment balance in the sinking fund is neither a tax-deductible expense nor is it an actual cash transaction.

TABLE 5.11

Cash Flow Timeline for Retrieval Vehicle (given loan of $28,000, capital acquisition −$28,000, salvage value at 60 months of $9,175, and monthly depreciation PMT$_{Dep}$ of −$314)

t (months)	Capital Costs ($) PMT$_{Loan}$	Sinking Fund ($) PMT$_{SF}$	Bal$_{SF}$	Capital Tax ($) Book Val	PMT$_{CT}$	Operating and Maintenance Costs ($) Exp$_{DLO}$	Exp$_{DLM}$	Exp$_{Matls}$	Exp$_{DOH}$	ΣFTD$_{Exp}$	Tax Table ($) F$_{Tax}$	ΣSTD$_{Exp}$	S$_{tax}$	NCF ($) Pretax	Posttax
(1)	(2)	(3)	(4)	(5)	(6)	(7)	(8)	(9)	(10)	(11)	(12)	(13)	(14)	(15)	(16)
0				28,000											
1		−688	688	27,686	−50	−34,253	−4,757	−1,111	−2,716	−43,201	16,848	−26,353	1,739	−43,525	−24,988
2		−688	1,381	27,373	−50	−34,253	−4,757	−1,116	−2,716	−43,206	16,850	−26,355	1,739	−43,530	−24,990
3	−315	−688	2,079	27,059	−49	−34,253	−4,757	−1,121	−2,716	−43,525	16,975	−26,550	1,752	−43,851	−25,172
4		−688	2,781	26,745	−48	−34,253	−4,757	−1,127	−2,716	−43,215	16,854	−26,361	1,740	−43,541	−24,996
5		−688	3,488	26,431	−48	−34,253	−4,757	−1,132	−2,716	−43,220	16,856	−26,364	1,740	−43,546	−24,998
6	−315	−688	4,201	26,118	−47	−34,253	−4,757	−1,137	−2,716	−43,540	16,980	−26,559	1,753	−43,867	−25,181
7		−688	4,918	25,804	−47	−34,823	−4,856	−1,143	−2,716	−43,898	17,120	−26,778	1,767	−44,226	−25,385
8		−688	5,640	25,490	−46	−34,823	−4,856	−1,148	−2,716	−43,903	17,122	−26,781	1,768	−44,231	−25,388
9	−315	−688	6,366	25,176	−46	−34,823	−4,856	−1,154	−2,716	−44,223	17,247	−26,976	1,780	−44,552	−25,570
10		−688	7,098	24,863	−45	−34,823	−4,856	−1,159	−2,716	−43,913	17,126	−26,787	1,768	−44,242	−25,393
11		−688	7,835	24,549	−45	−34,823	−4,856	−1,164	−2,716	−43,918	17,128	−26,790	1,768	−44,248	−25,396
12	−315	−688	8,577	24,235	−44	−34,823	−4,856	−1,170	−2,716	−44,238	17,253	−26,985	1,781	−44,568	−25,578
13		−688	9,324	23,921	−43	−35,403	−4,957	−1,176	−2,761	−44,654	17,415	−27,239	1,798	−44,985	−25,815
14		−688	10,077	23,608	−43	−35,403	−4,957	−1,181	−2,761	−44,659	17,417	−27,242	1,798	−44,990	−25,818
15	−315	−688	10,834	23,294	−42	−35,403	−4,957	−1,187	−2,761	−44,979	17,542	−27,437	1,811	−45,311	−26,001
16		−688	11,597	22,980	−42	−35,403	−4,957	−1,192	−2,761	−44,669	17,421	−27,248	1,798	−45,001	−25,824
17		−688	12,365	22,666	−41	−35,403	−4,957	−1,198	−2,761	−44,674	17,423	−27,251	1,799	−45,007	−25,827
18	−315	−688	13,138	22,353	−41	−35,403	−4,957	−1,204	−2,761	−44,994	17,548	−27,446	1,811	−45,328	−26,009

19		−688	13,916	22,039	−40	−35,992	−5,060	−1,209	−2,761	−45,376	17,697	−27,680	1,827	−45,711	−26,227
20	−315	−688	14,700	21,725	−39	−35,992	−5,060	−1,215	−2,761	−45,381	17,699	−27,683	1,827	−45,716	−26,230
21		−688	15,489	21,411	−39	−35,992	−5,060	−1,221	−2,761	−45,702	17,824	−27,878	1,840	−46,037	−26,412
22		−688	16,284	21,098	−38	−35,992	−5,060	−1,226	−2,761	−45,392	17,703	−27,689	1,827	−45,728	−26,236
23	−315	−688	17,084	20,784	−38	−35,992	−5,060	−1,232	−2,761	−45,397	17,705	−27,692	1,828	−45,734	−26,239
24		−688	17,890	20,470	−37	−35,992	−5,060	−1,238	−2,761	−45,717	17,830	−27,888	1,841	−46,054	−26,421
25		−688	18,701	20,156	−37	−36,592	−5,155	−1,244	−2,807	−46,158	18,002	−28,156	1,858	−46,496	−26,672
26		−688	19,518	19,843	−36	−36,592	−5,155	−1,250	−2,807	−46,163	18,004	−28,160	1,859	−46,502	−26,675
27	−315	−688	20,340	19,529	−36	−36,592	−5,165	−1,256	−2,807	−46,484	18,129	−28,355	1,871	−46,822	−26,858
28		−688	21,168	19,215	−35	−36,592	−5,165	−1,262	−2,807	−46,174	18,008	−28,166	1,859	−46,513	−26,682
29		−688	22,002	18,901	−34	−36,592	−5,165	−1,268	−2,807	−46,179	18,010	−28,169	1,859	−46,519	−26,685
30	−315	−688	22,841	18,588	−34	−36,592	−5,165	−1,274	−2,807	−46,500	18,135	−28,365	1,872	−46,840	−26,867
31		−688	23,686	18,274	−33	−37,201	−5,272	−1,280	−2,807	−46,907	18,294	−28,613	1,888	−47,248	−27,099
32		−688	24,537	17,960	−33	−37,201	−5,272	−1,286	−2,807	−46,912	18,296	−28,616	1,889	−47,254	−27,102
33	−315	−688	25,394	17,646	−32	−37,201	−5,272	−1,292	−2,807	−47,233	18,421	−28,812	1,902	−47,575	−27,285
34		−688	26,257	17,333	−32	−37,201	−5,272	−1,298	−2,807	−46,923	18,300	−28,623	1,889	−47,266	−27,108
35		−688	27,125	17,019	−31	−37,201	−5,272	−1,304	−2,307	−46,929	18,302	−28,627	1,889	−47,272	−27,112
36	−315	−688	28,000	16,705	−30	−37,201	−5,272	−1,310	−2,307	−47,249	18,427	−28,822	1,902	−47,593	−27,294
37				16,391	−30	−37,820	−5,381	−1,316	−2,854	−47,716	18,609	−29,106	1,921	−47,372	−26,872
38				16,078	−29	−37,820	−5,381	−1,323	−2,854	−47,721	18,611	−29,110	1,921	−47,378	−26,875
39				15,764	−29	−37,820	−5,381	−1,329	−2,854	−47,727	18,613	−29,113	1,921	−47,384	−26,878
40				15,450	−28	−37,820	−5,381	−1,335	−2,854	−47,733	18,616	−29,117	1,922	−47,391	−26,881
41				15,136	−28	−37,820	−5,381	−1,341	−2,854	−47,738	18,618	−29,120	1,922	−47,397	−26,885
42				14,823	−27	−37,820	−5,381	−1,348	−2,854	−47,744	18,620	−29,124	1,922	−47,403	−26,888
43				14,509	−27	−38,450	−5,493	−1,354	−2,854	−48,491	18,912	−29,580	1,952	−48,151	−27,314
44				14,195	−26	−38,450	−5,493	−1,360	−2,854	−48,497	18,914	−29,583	1,952	−48,157	−27,317

continued

TABLE 5.11 (continued)

Cash Flow Timeline for Retrieval Vehicle (given loan of $28,000, capital acquisition −$28,000, salvage value at 60 months of $9,175, and monthly depreciation PMT_{Dep} of −$314)

t (months)	Capital Costs ($) PMT_{Loan}	Sinking Fund ($) PMT'_{SF}	Bal_{SF}	Capital Tax ($) BookVal	PMT_{CT}	Operating and Maintenance Costs ($) Exp_{DLO}	Exp_{DLM}	Exp_{Matls}	Exp_{DOH}	ΣFTD_{Exp}	Tax Table ($) F_{Tax}	ΣSTD_{Exp}	S_{tax}	NCF ($) Pretax	Posttax
(1)	(2)	(3)	(4)	(5)	(6)	(7)	(8)	(9)	(10)	(11)	(12)	(13)	(14)	(15)	(16)
45				13,881	−25	−38,450	−5,493	−1,367	−2,854	−48,503	18,916	−29,587	1,953	−48,164	−27,320
46				13,568	−25	−38,450	−5,493	−1,373	−2,854	−48,509	18,918	−29,590	1,953	−48,170	−27,324
47				13,254	−24	−38,450	−5,493	−1,380	−2,854	−48,515	18,921	−29,594	1,953	−48,177	−27,327
48				12,940	−24	−38,450	−5,493	−1,386	−2,854	−48,521	18,923	−29,598	1,953	−48,183	−27,330
49				12,626	−23	−39,090	−5,607	−1,393	−2,902	−49,329	19,238	−30,090	1,986	−48,992	−27,791
50				12,313	−23	−39,090	−5,607	−1,399	−2,902	−49,335	19,240	−30,094	1,986	−48,998	−27,794
51				11,999	−22	−39,090	−5,607	−1,406	−2,902	−49,341	19,243	−30,098	1,986	−49,005	−27,798
52				11,685	−21	−39,090	−5,607	−1,413	−2,902	−49,347	19,245	−30,101	1,987	−49,011	−27,801
53				11,371	−21	−39,090	−5,607	−1,419	−2,902	−49,353	19,248	−30,105	1,987	−49,018	−27,805
54				11,058	−20	−39,090	−5,607	−1,426	−2,902	−49,359	19,250	−30,109	1,987	−49,025	−27,808
55				10,744	−20	−39,741	−5,723	−1,433	−2,902	−50,132	19,552	−30,581	2,018	−49,799	−28,249
56				10,430	−19	−39,741	−5,723	−1,440	−2,902	−50,139	19,554	−30,584	2,019	−49,806	−28,252
57				10,116	−19	−39,741	−5,723	−1,446	−2,902	−50,145	19,556	−30,588	2,019	−49,812	−28,256
58				9,803	−18	−39,741	−5,723	−1,453	−2,902	−50,151	19,559	−30,592	2,019	−49,819	−28,259
59				9,489	−18	−39,741	−5,723	−1,460	−2,902	−50,157	19,561	−30,596	2,019	−49,826	−28,263
60				9,175	−17	−39,741	−5,723	−1,467	−2,902	−40,989	15,986	−25,003	1,650	−49,833	−23,039

- The monthly depreciation expense through the end of the fifth year, month 60 (PMT_{Dep}), is given in the table title. Depreciation expenses are tax-deductible but are not an actual cash transaction. Depreciation is an accountant's contrivance used to write off a capital asset.

- Column 5 is an internal working table that calculates the book value (BookVal) of the retrieval vehicle through the end of its useful life at month 36. The book value decreases to equal the salvage value in the last depreciation period. The book value is neither a tax-deductible expense nor is it an actual transaction. It does calculate the capital valuation used to compute the capital tax expense.

- Column 6 records the capital tax expense over the useful life of the asset (PMT_{CT}). The capital tax expense is a tax-deductible expense and an actual cash transaction.

Operating and Maintenance Costs

All operating and maintenance costs are tax-deductible expenses and are actual cash transactions.

- Column 7 records the estimation of direct labor expenses for retrieval vehicle operations starting with the baseline annual labor expense divided by 12 months (EXP_{DLO}). The initial estimate for the monthly direct labor expense is a uniform recurring amount through the end of the sixth month. The semiannual direct labor cost escalator is applied to calculate the estimated direct labor expenses for the next six months. The semiannual direct labor cost escalator is applied every six months through the end of the evaluation period for the retrieval vehicle, month 60.

- Column 8 records the estimation of direct labor expenses for retrieval vehicle maintenance (EXP_{DLM}). The initial estimate for the monthly direct labor expense is calculated in the same manner for direct labor expenses as for retrieval vehicle operations. The frequency of the labor cost escalator is identical, but the rate is different. Had the rates been equal, the two labor expense categories can be consolidated into a single column.

- Column 9 records the estimation of direct materials expenses for retrieval vehicle operations and maintenance (EXP_{Matls}). The frequency of the direct materials cost escalator is monthly. Direct materials expenses are recurring amounts but not uniform.

- Column 10 records the estimation for direct overhead for operations and maintenance of the retrieval vehicle (EXP_{DOH}). The frequency of the direct overhead cost escalator is annual. Direct overhead expenses are uniform recurring amounts in each year.

- Column 11 is an internal worksheet that sums the federal tax-deductible expenses. It is the basis for calculating the federal tax credit for tax-deductible expenses.
- Column 12 is the federal tax credit. The federal tax credit adjusts the magnitude of the expenses for calculating the net cash flow for a time period and is an actual cash transaction. Although it is calculated monthly in this cash flow worksheet, the actual frequency of tax reporting for organizations is quarterly. The federal tax credit is one example of the distinction between a cash flow timeline used for engineering economic analysis and accounting practices. The frequency of calculating the federal tax credit is driven by the time units for the timeline.
- Column 13 is an internal worksheet that sums the state tax-deductible expenses. Typically state tax-deductible expenses include federal tax expenses. It is the basis for calculating the state tax credit for tax-deductible expenses.
- Column 14 is the state tax credit. It conforms to the calculation approach used for the federal tax credit.

Net Cash Flow

- Column 15 is the net cash flow for all pretax expenses. It is the sum of all actual cash transactions before the application of tax credits. It includes both tax-deductible and non–tax-deductible expenses.
- Column 16 is the net cash flow (NCF) that applies all tax credits to the pretax expenses. This column would be equal to Column 15 if organizations did not incur tax obligations.

The distinction between these two columns is critical to logistical economic analysis. If the reliability engineering analysis combined with logistical economic analysis did not address the impact of tax credits then the cost estimations in their analyses would be overstated. Too often, engineers equate the price tag for labor and materials as the actual cost to the organization.

Hourly Rate Estimation for the Retrieval Vehicle

The hourly rate estimation for the retrieval vehicle is calculated from the posttax net cash flow column from the cash flow timeline. The sum of the posttax net cash flows for each month in year 1 is divided by the utilization factor for the retrieval vehicle. This provides us with an hourly rate estimation of $29. The same procedure allows us to estimate the hourly rate for the retrieval vehicle in the second through the fifth year, as shown in Table 5.12.

We observe that the increase in the hourly rate for the retrieval vehicle is a composite of all of the expenses and their respective escalation factors.

TABLE 5.12

Cost Estimations for Retrieval Vehicle

	Year 1	Year 2	Year 3	Year 4	Year 5
Annual cost ($)	−303,036	−313,060	−323,439	−325,211	−331,114
O&M cost ($/hr)	36	37	39	39	39

Example: Facility Maintenance Bay

The objective of logistics economic analysis is to calculate the facility cost per bay per hour.

The capital investment in a facility is typically financed with a mortgage. Mortgage loans are repaid using the capital recovery in which a repayment schedule is comprised of equal payments. Each payment includes payment of interest on the unpaid balance and payment on the loan balance.

The following economic information is provided:

- The present amount of the capital investment, $P = -\$9,550,000$
- The mortgage interest rate, $r = 5.13\%$ APR (the interest rate per payment period, $i = r/m = 0.004271$ per month)
- The duration of the mortgage loan, $N = 10$ years
- The payment frequency, $m = 12$ months/year (the total number of payments, $n = mN = 120$ months)

The loan payment per period is calculated in Excel, PMT = −$101,887/month. The excel approach uses the following paste function:

$$PMT(rate, nper, pv, fv) = PMT(0.004271, 120, -\$9,550,000)$$

where "rate" is the interest rate per payment period, "nper" is the total number of payment periods, and "pv" is the present amount of the loan (the "fv" argument is left blank).

Maintenance events are performed 24 hours/day, 30 days/month = 720 hours/month (engineering economic analysis assumes a 30 day month). The hourly capital recovery amount for the facility is calculated as the loan payment/month divided by the number of hours/month, −$101,887/720 = −$141.50/hour.

The maintenance facility has 12 bays. The facility cost per bay per hour is calculated as the hourly capital recovery amount for the facility divided by the number of maintenance bays, −$141.50/12 = −$11.79/bay hour.

The facility cost per bay per hour is a uniform recurring amount that does not increase over the duration of the mortgage loan and is an actual cash transaction. The procedure is illustrated in Table 5.13.

TABLE 5.13

Facility Fixed Cost Estimation for Facility
Bay Hours

Capitalized investment, P ($)	9,550,000
Cost of capital, r (% APR)	5.13
N (yr)	10
m (mon/pp)	12
i (per month)	0.004271
n (months)	120
Payment/month PMT ($/month)	101,877
Scheduled hours (hr/month)	$720 = 24 \times 30$
Facility cost ($/hr)	141.50/hr
Number of bays	12
Facility cost ($/bay hr)	11.79

Lost Opportunity Costs

Lost opportunity costs are the forfeited revenue caused by the unavailability of a system due to unscheduled corrective maintenance and scheduled preventive maintenance. Consider an organization that generates $465,000,000 in revenue from operations. The organization operates 18 production vehicles. The planned, or budgeted, revenue per vehicle per year is the revenue divided by the number of vehicles, $465,000,000/18 = $25,833,333/vehicle year. The organization operates 8 hours/shift, 3 shifts/day, 7 days/week, 50 weeks/year = 8,400 hours/year. Each vehicle is scheduled for 30 minutes of servicing between shifts. That yields a utilization rate of 0.9375 (7.5 hr/8 hr). The planned annual production hours per vehicle is calculated as the product of total operating hours and the utilization factor, 8,400 hr/yr × 0.9375 = 7,875 hr/yr, as shown in Table 5.14.

The lost opportunity cost per hour of downtime for each vehicle is calculated as the planned annual production revenue divided by the planned annual production hours per vehicle, $25,833,333/7,875 = –$3,280/downtime hour.

TABLE 5.14

Planned Annual Production Hours

Utilization rate = 7.5/8	0.9375
total hours (per shift)	8
service hours (per shift)	0.5
production hours (per shift)	7.5
Total production hours (hr/yr)	8,400
(50 wks/yr)(7 d/wk)(8 hrs/shift)(3 shifts/d)	
Planned annual production hours/vehicle	7,875

TABLE 5.15

Lost Opportunity Hourly Cost Estimation per System

Annual production revenue goal ($)	465,000,000
Production vehicles	18
Planned annual production revenue ($/vehicle)	25,833,333
Planned annual production hours (per vehicle)	7,875
Lost opportunity cost/hr downtime ($)	3,280.42

Lost opportunity costs/hour are not tax deductible expenses, nor are they actual transactions, but the impact of the lost revenue is typically greater than the hourly expensed capital, operating expenses, and maintenance expenses (Table 5.15).

Reliability failure, repair, and logistics downtime math models determine the frequency of the occurrence of system downing events and the duration of the total downtime that causes lost opportunity costs. The transition from unscheduled corrective maintenance actions to scheduled preventive maintenance actions minimizes lost opportunity costs.

6

Life-Cycle Economic Analysis

This chapter puts all of the pieces of the puzzle together and enables managers and engineers to apply economic analysis to reliability investigations and logistical information to perform life-cycle economic analysis for system design and sustainment. Two approaches are presented:

1. Deterministic
2. Simulation

Deterministic analysis uses point estimates of means for reliability, i.e., operational availability, time to failure, time to repair, etc. Simulation is introduced to evaluate reliability functions over time and allows for repetition of the failure–maintenance cycle. Most important, simulation investigates what happens to a system after the first part fails, the second, and so on. Deterministic analysis cannot do that. This chapter will use information provided by reliability simulation as inputs to life-cycle economic analysis.

The scope of the chapter addresses three applications of life-cycle economic analysis for system design and sustainment:

- Part selection
  - Design-for-reliability analysis investigates alternatives for make–buy and purchase options.
  - Reliability-based sustainment investigates alternatives for original equipment manufacturer (OEM) and third party vendors for spare part options.
- Preventive maintenance
  - Design-for-maintainability analysis investigates recommended scheduled maintenance intervals.
  - Reliability-based sustainment investigates actual requirements for scheduled maintenance intervals.
- Spare parts strategy
  - Design-for-maintainability analysis investigates recommended part order quantities and frequency.
  - Reliability-based sustainment investigates actual requirements part order quantities and frequency.

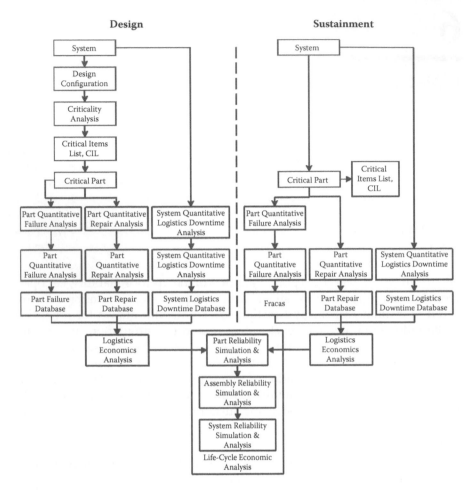

FIGURE 6.1
Sustainment life-cycle economic analysis flowchart.

The fundamental principle that applies to all life-cycle economic analyses is that availability is inversely related to total downtime for parts: availability decreases when total downtime increases and availability increases when total downtime decreases. Recall that all sustainment costs are caused by part failure, availability is a function of part time to failure, and total downtime is a measure of mitigation for part failure.

The design and sustainment paths to determine the impact of part selection, preventive maintenance, and spare parts strategy decisions on availability and total downtime is summarized in Figure 6.1.

- Life-cycle economic analyses begin with identification of critical parts that warrant investment in reliability analyses. Design and sustainment processes differ:

- Design requires part criticality analyses for candidate parts identified through a functional fault tree analysis to create a rank-ordered critical items list from which critical parts are selected.
- Sustainment is based on a critical part's maintenance demand and create the critical items list.
- Reliability analyses include investigations and development of math models for time to failure, time to repair, and logistics downtime (LDT).
  - Design and sustainment failure math models are developed from time-to-failure data. Design failure math models require a qualitative failure analysis to design the failure experiment to acquire time-to-failure data. Sustainment performs a qualitative failure analysis to implement a failure report, root cause analysis, and corrective action system. Sustainment acquires time-to-failure data from historical maintenance records.
  - Design and sustainment repair math models are developed from time-to-repair data. Design repair math models require a qualitative repair analysis to design the repair experiment to acquire time-to-repair data. Sustainment acquires time-to-repair data from historical maintenance records.
  - Design and sustainment logistics downtime math models are developed from subject matter Delphi surveys. Design logistics downtime math models require a qualitative failure analysis to determine the parameters of the logistics downtime math models. Sustainment acquires logistics downtime data from historical maintenance records.
- Logistics economics analysis investigates logistical resources and their allocations to implement maintenance actions (prerepair logistics, repair, and postrepair logistics).
  - Design relies on analysis to determine estimates for logistical resources.
- Sustainment relies on current data from management information systems to calculate estimates for logistical resources.

The paths converge at this point. Reliability and logistics economic information is applied to critical parts to make part selection decisions and evaluate the decisions on the part life-cycle economic analysis (Figure 6.1). Parts are integrated into assembly design configurations to evaluate part selection, preventive maintenance, and spare part strategy decisions on the assembly life-cycle economic analysis. Assemblies are integrated into the system design configuration to evaluate all part selection, preventive maintenance, and spare part strategy decisions on the system life-cycle economic analysis.

Part Selection

The part selection life-cycle economic analysis procedure is illustrated by an example.

- Design analysis allocates two reliability requirements to the part: mean time between failure (MTBF), at least 300 hours, and A_O, at least 96%.
- Sustainment evaluates the OEM part and finds that it has an MTBF = 300 hours, and an A_O = 96%.

An obvious requirement is to select the lowest cost part alternative that meets the reliability requirements. The requirements are summarized in Table 6.1.

Of all of the parts investigated, five parts are identified that meet the requirements, Alpha, Beta, Charley, Delta, and Echo. The deterministic point estimates for the mean reliability metrics and the vendors' prices are provided in Table 6.2.

Too many organizations make the decision to select part Beta at this juncture based on vendor cost. These same organizations make no further use of the wealth of information that was acquired to perform the reliability investigations to characterize the MTBF and A_O.

Is part Beta actually the lowest cost part selection? Certainly it has the lowest acquisition cost, but what is the part selection impact on the system life-cycle cost to the system user (the organization that will operate and maintain the system)? Let us first apply the deterministic approach to evaluating the life-cycle economic analysis, to be followed by the simulation approach.

TABLE 6.1

Requirements Table

MTBF (hr) > 300
A_O > 0.96
Lowest cost

TABLE 6.2

Part Summary Data

Part	MTBF	A_O	Cost ($)
Alpha	422.8	0.975	5,333
Beta	310.0	0.964	3,072
Charley	473.6	0.972	4,667
Delta	688.0	0.983	6,188
Echo	417.4	0.972	3,645

TABLE 6.3

Part Alpha Failure Repair Logistics Reliability Analysis

	Failure Math Model			Repair Math Model			System		
	η	β	γ	η	β	γ			
	35.2	6.67	390	0.56	2.25	3.15	Λ_{prerep}	$\Lambda_{postrep}$	A_O
μ			422.84			3.65	5.528	1.509	0.975
Confidence limit			412.56			4.06	8.474	2.025	0.966

Deterministic Reliability Analysis

Reliability investigations were performed to determine whether the parts achieved the required reliability experiments to acquire data to fit the part failure, repair, and logistics downtime math model. Math model parameters for part Alpha are summarized in Table 6.3.

Reliability math models yield the point estimates for the mean (μ), time to failure, time to repair, prerepair logistics downtime and postrepair logistics downtime. Deterministic reliability math models also yield the lower confidence limit (CL) for time to failure, and the upper confidence limits for time to repair, prerepair logistics downtime and postrepair logistics downtime. The deterministic point estimate for A_O and its lower confidence limit is calculated from means and confidence limits for time to failure, time to repair, prerepair logistics downtime, and postrepair logistics downtime.

The math model parameters for parts Beta, Charley, Delta, and Echo, are summarized in Tables 6.4–6.7. Also included in the tables are the point

TABLE 6.4

Part Baker Failure Repair Logistics Reliability Analysis

	Failure Math Model			Repair Math Model			System		
	η	β	γ	η	β	γ			
	16.8	2.97	295	0.91	2.46	3.65	Λ_{prerep}	$\Lambda_{postrep}$	A_O
μ			310.00			4.46	5.528	1.509	0.964
Confidence limit			301.17			5.07	8.474	2.025	0.951

TABLE 6.5

Part Charley Failure Repair Logistics Reliability Analysis

	Failure Math Model			Repair Math Model			System		
	η	β	γ	η	β	γ			
	25.3	6.67	450	0.55	2.25	6.25	Λ_{prerep}	$\Lambda_{postrep}$	A_O
μ			473.61			6.74	5.528	1.509	0.972
Confidence limit			466.25			7.15	8.474	2.025	0.964

TABLE 6.6

Part Delta Failure Repair Logistics Reliability Analysis

	Failure Math Model			Repair Math Model					
	η	β	γ	η	β	γ	System		
	48.9	7.93	642	0.67	3.05	4.33	Λ_{prerep}	$\Lambda_{postrep}$	A_O
μ			688.03			4.93	5.528	1.509	0.983
Confidence limit			675.62			5.29	8.474	2.025	0.977

TABLE 6.7

Part Echo Failure Repair Logistics Reliability Analysis

	Failure Math Model			Repair Math Model					
	η	β	γ	η	β	γ	System		
	18.5	2.15	401	1.55	2.57	3.67	Λ_{prerep}	$\Lambda_{postrep}$	A_O
μ			417.38			5.05	5.528	1.509	0.972
Confidence limit			405.65			6.05	8.474	2.025	0.961

estimates for MTBF, mean time to repair (MTTR), operational availability, and their confidence limits.

Each table provides the reliability multipliers that will be applied to the labor, materials and overhead costs to calculate the present, uniform recurring, and future amounts that will determine the net cash flow for maintenance events that occur as a part fails.

Labor Costs

We begin to calculate the hourly labor costs estimated to perform prerepair logistics, repair, and postrepair logistics events by identifying the labor skills categories required. Labor skills categories are identified by the qualitative repair and logistics analysis performed in design for maintainability and from best practices developed in sustainment for this critical part. Labor skills categories and the respective hourly labor costs estimates are summarized in Table 6.8.

The labor skills category allocations are determined by the qualitative repair and logistics analysis performed in design for maintainability and from best practices developed in sustainment. For example, 10% of the supervisor labor category is allocated to prerepair logistics events, 33% is allocated to repair events, and 25% is allocated to postrepair for this critical part.

Hourly labor costs are estimated for prerepair logistics, repair, and postrepair logistics events by labor skills category. For example, the hourly labor costs for the supervisor for the prerepair logistics events is equal to the product of the hourly labor cost and allocation: $130.70 times 10% is equal to $13.07. The remaining hourly labor costs for prerepair logistics events are calculated

TABLE 6.8

Labor Allocation and Cost Table

Labor Cost Estimation		Labor Allocation (per hr)			Labor Cost ($/hr)		
Labor Category	Rate ($/hr)	Pre-repair	Repair	Post-repair	Pre-repair	Repair	Post-repair
Supervisor	130.70	0.1	0.33	0.25	13	43	33
Master tech 1	111.09	0.25	1.5	0	28	167	0
Master tech 2	98.02	0	2	0.33	0	196	32
Journeyman 1	84.95	0.5	1	1	42	85	85
Journeyman 2	65.35	0	2	0	0	131	0
Helper	52.28	1	4	1	52	209	52
Maintenance planner	117.63	0.33	0.25	0.5	39	29	59
Supply clerk	58.81	0.25	0	0.25	15	0	15
					189	860	276

in the same way. The hourly labor cost for prerepair logistics events is equal to the sum of the hourly labor costs by skills category, in this example $189. The end result is the hourly labor costs for prerepair logistics, repair, and postrepair logistics events. The three-step procedure for tabulating and calculating the respective costs is presented in Table 6.8.

Materials Cost for Spare Parts

The vendor price was presented in the initial analysis of the part alternatives. It is not the cost of the part to the organization regardless of whether the designer or sustainer acquires the part, as was observed in the logistical cost analysis. The vendor price is a raw cost that must be burdened by the organizations cost of doing business to acquire the part. The actual cost of the spare part to the organization is tabulated in the cost estimation table (Table 6.9).

TABLE 6.9

Materials Cost Table

		Cost Estimation	
Vendor Price ($)		Materials	Unit Cost ($)
Part A	5,333	Spare Part A	7,786
Part B	3,072	Spare Part B	4,485
Part C	4,667	Spare Part C	6,814
Part D	6,188	Spare Part D	9,034
Part E	3,645	Spare Part E	5,322
Hardware	573	Hardware kit	837
		Inventory carrying cost ($/hr)	2.35

Often the vendor cost that is tabulated is incomplete when spare part acquisition investigations are performed. Observe that the spare part requires hardware to include fasteners, interfaces, etc. As for the spare part, the raw cost is the cost of the hardware plus the burden to the organization of acquiring the hardware. In this example, the same hardware is required regardless of the selection of the spare part vendor.

Overhead

Overhead costs are costs that apply specifically to the part maintenance action, broken down into prerepair logistics, repair, and postrepair logistics events (Table 6.10). Logistical cost analysis determined that the hourly cost to the organization for the use of the retrieval vehicle is $47 per hour. The retrieval vehicle is used during prerepair logistics downtime events. The overhead crane and facility bay are hourly overhead costs that are applied during part repair events.

Part Selection Life-Cycle Economic Analysis

The procedure for performing a part selection life-cycle economic analysis applies as a special case for the cash flow timeline. The decision criterion for the part selection life-cycle economic analysis is the equivalent uniform recurring amount. The deterministic approach evaluates each part for the point estimate of the mean time between failure. The point estimate of the MTBF will be the duration of the economic analysis evaluation period for each part.

Another distinction for a part selection life-cycle economic analysis is the selection of timeline units. All reliability and logistical economic metrics are stated in hours; therefore the time units will be hours.

Part option cost events are estimated for each part as shown in Table 6.11. The procedure for determining the part option cost events requires estimation of present amounts, uniform recurring amounts, and future amounts associated with the labor, materials, and overhead costs.

TABLE 6.10

Direct Overhead Cost Estimation Table

Retrieval vehicle ($/hr)	47
Overhead crane ($/hr)	13
Facility bay ($/hr)	340

TABLE 6.11

Part Option Cost Table

	Alpha	Baker	Charley	Delta	Echo
P: Present amounts					
Part cost ($)	7,786	4,485	6,814	9,034	5,322
Hardware ($)	837	837	837	837	837
A: Uniform recurring amounts					
ICC ($)	2	2	2	2	2
F: Future amounts					
Prerepair logistics cost events					
A_{prerep}	5.53	5.53	5.53	5.53	5.53
Labor ($)	1,045	1,045	1,045	1,045	1,045
Cost ($/hr)	189	189	189	189	189
Overhead					
Retrieval vehicle ($)	199	199	199	199	199
Cost ($/hr)	36	36	36	36	36
Repair cost events					
MTTR	3.65	4.46	6.74	4.93	5.05
Labor ($)	3,135	3,833	5,794	4,239	4,340
Cost ($/hr)	860	860	860	860	860
Overhead					
Crane ($)	14	18	27	19	20
Cost ($/hr)	4	4	4	4	4
Facility/Bay ($)	43	53	79	58	60
Cost ($/hr)	12	12	12	12	12
Postrepair logistics cost events					
$A_{postRep}$	1.51	1.51	1.51	1.51	1.51
Labor ($)	416	416	416	416	416
Cost ($/hr)	276	276	276	276	276

Present Amount

Part selection life-cycle cost analysis makes the assumption that a spare part is acquired upon replacement of that part. Therefore the evaluation cycle begins when the part is installed and a replacement part is acquired. The present amount includes the costs to the organization for the spare part and hardware.

Uniform Recurring Amounts

The most common uniform recurring amount for the part selection economic analysis is inventory carrying costs. Inventory carrying costs are allocated to the part while it is in storage awaiting use.

Future Amounts

Future amounts are calculated for each of the maintenance events, prerepair logistics downtime, repair, and postrepair logistics downtime. Applicable labor, materials, and overhead costs are allocated to each of these events. The future amounts for prerepair logistics cost events include labor and overhead. There are no material costs for prerepair logistics downtime.

For convenience, the point estimate of the mean prerepair logistics downtime is entered into the table (Table 6.11). The future amount for labor is equal to the product of the mean prerepair logistics downtime and the hourly cost for the labor skills categories. Note that the duration and hourly rate are equal for all five parts. There is a misunderstood conception in performing engineering economic analysis that costs that are equal for all alternatives may be excluded from the analysis. This is not the case for part selection, because the cost of an event occurs at different times for each part. Only when the costs and occurrences are equal may they be eliminated. However the retrieval vehicle is an overhead cost. The overhead costs for the retrieval vehicle are equal to the product of the mean prerepair logistics downtime and the hourly cost of the retrieval vehicle.

- Future amounts for repair cost events include labor and overhead. Materials cost, spare parts, and hardware costs are accounted for as a present amount.
- Future amounts for the labor cost are the products of the mean time to repair and the hourly labor cost for the crane and the facility bay. Each overhead cost is equal to the MTTR and its respective hourly cost to the organization.
- Future amount for the postrepair logistics events is labor only.
- Future amounts for postrepair logistics labor is equal to the mean postrepair logistics downtime and the hourly cost of labor.

Economic Factors

Engineering economic analyses calculate the equivalent values based on the organization's internal rate of return (IRR), expressed as in annual percentage rate (APR). The interest rate per compounding period is equal to the APR divided by the total number of compounding periods in a year. As shown in the logistical cost analysis, the organization operates 8,400 hours per year. The interest rate per compounding period is equal to 0.0000149/hr.

Part Option Economic Analysis for Expense to Costs

The present, uniform recurring, and future amounts represent costs that are tracked by the management information system and reported in accounting

TABLE 6.12

Part Option Economic Analysis: Expensed Costs[a]

	Alpha	Baker	Charley	Delta	Echo
n	423	310	474	688	417
P ($)	−8,623	−5,322	−7,650	−9,871	−6,158
A ($)	−2	−2	−2	−2	−2
PV[A] ($)	−991	−727	−1,109	−1,609	−978
F ($)	−4,854	−5,564	−7,561	−5,977	−6,080
PV[F] ($)	−4,823	−5,539	−7,508	−5,916	−6,042
NPV ($)	−14,437	−11,587	−16,267	−17,396	−13,179
A_{Equiv} ($)	−34.25	−37.46	−34.47	−25.41	−31.67
A_O	0.9754	0.9642	0.9717	0.9829	0.9719

[a] Economic factors: $r_{\text{IRR}} = 12.50\%$; $m = 8,400$, and $i_{\text{IRR}} = 1.49 \times 10^{-5}$.

records. Each cost is viewed by the organization as an expense that is an actual cash flow. The procedure for performing the part option economic analysis is summarized in Table 6.12. Each part is assigned a column.

- The first row, n, is the total number of compounding periods in hours that is equal to the point estimate of the MTBF.
- P is the sum of the present amounts for each part from the part option cost events table.
- A is the sum of the uniform recurring amounts for each part.
- PV[A] is the calculation for the present equivalent value of the uniform recurring amounts for each part.
- F is the sum of the future amounts for each part.
- PV[F] is the present equivalent value of the future amounts for each part.
- NPV is the sum of the present amount and the equivalent present values, PV + PVA + PVF for each part.
- A_{Equiv} is the equivalent uniform recurring amount per hour for the net present value (NPV) for each part.

Part Baker has the lowest net present value, and part Delta has the highest net present value. These two net present values, along with the other three parts, cannot be evaluated head to head because they are based on different evaluation durations. The net present value for part Baker is evaluated for 310 evaluation periods, and part Delta is evaluated for 688 evaluation periods.

Observe that the relationship between the equivalent uniform recurring amount per hour for part Baker, the lowest raw cost and minimum net present value, and part Delta, the highest raw cost and maximum net present value, are reversed. Part Delta, with an equivalent uniform recurring amount of $25 per hour, is the lowest life-cycle cost option of the five parts. Part Baker, with an equivalent uniform recurring amount of $37 per hour, is the highest life-cycle cost option of the five parts.

Part Option Economic Analysis for Lost Opportunity Cost

The previous economic analysis was performed for expense costs associated with present, recurring uniform, and future amounts. Organizations incur lost opportunity costs resulting from loss of the productive capacity of the system from critical part failure. The key metric for determination of lost opportunity cost is the point estimate of the mean total downtime (MDT). From the logistical cost analysis, the lost opportunity cost per hour for the system is $3,280. The lost opportunity cost for part failure events is equal to the product of the lost opportunity cost per hour and the mean total downtime. The occurrence of the lost opportunity cost is the point estimate of the MTBF for each part. Lost opportunity cost is treated as a future amount. The present value of the lost opportunity cost is calculated for the interest rate per compounding period, the number of compounding periods, and the MTBF for each part. Notice that the lowest lost opportunity cost is provided by part Delta, and the highest is provided by part Baker. Table 6.13 summarizes the part option economic analysis for the lost opportunity costs associated with each of the five parts and their equivalent uniform recurring cost for each present value.

The part option economic analysis can be effectively communicated to managers by plotting the equivalent uniform recurring amounts for expense and lost opportunity costs as shown in Figure 6.2. Not only does it graphically illustrate part Delta, which has the lowest life-cycle cost to the system user, it also shows the relative magnitudes and relationships between the five parts. It is readily apparent that part Delta incurs roughly half the life-cycle cost of part Baker.

TABLE 6.13

Part Option Economic Analysis: Lost Opportunity Costs

	Alpha	Baker	Charley	Delta	Echo
n	423	310	474	688	417
L_{Op} ($)	−35,045	−37,706	−45,185	−39,253	−39,638
MDT	10.68	11.49	13.77	11.97	12.08
Cost ($/hr)	−3,280	−3,280	−3,280	−3,280	−3,280
PV[L_{Op}] ($)	−34,825	−37,532	−44,868	−38,853	−39,393
A_{Equiv} ($)	−83	−121	−95	−57	−95

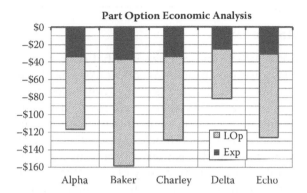

FIGURE 6.2
Equivalent uniform recurring amounts for expense and lost opportunity costs.

Cautionary Note

Many agree that it is intuitively apparent that high cost parts are more reliable and last longer than low cost parts. One can indeed drive the life-cycle cost of a system down through the selection of such parts and in doing so make the acquisition cost to the user uncompetitively high. An interesting study was published in a Harvard business review almost 50 years ago that investigated the life-cycle costs of ownership of the dominant automobiles that have been in service in the first half of the twentieth century. Many were startled to find that the Rolls-Royce provided the lowest life-cycle cost of ownership of any vehicle. Unfortunately, very few people could afford a Rolls-Royce. Engineering economics analysis provides information to managers and engineers that must be tempered by judgment. The numbers by themselves do not make a decision. The efficacy of this procedure is predicated on the criticality analysis that was performed by the organization that determined that a specific part warranted the investment in evaluating the reliability and economic analyses to make the best decision for the organization. Just as mindlessly seeking to find the lowest raw cost for a part is an extreme example of a bad approach, letting a life-cycle economic analysis decide the part selection absent judgment approaches the opposite extreme. Also recall that all reliability and economic metrics are estimates based on sampling and sampling error. The objective of this analysis is to provide the managing engineer with information that minimizes the uncertainty that applies to all decisions.

Preventive Maintenance

Preventive maintenance in an organization that operates in sustained systems is a continually evolving process over time. Assume that an organization employs a maintenance practice that is exclusively based on performing corrective maintenance to restore systems to functionality following unscheduled downing events resulting from critical part failures. Implementation of preventive maintenance is an incremental application of reliability and economic life-cycle analyses that transitions a part from corrective maintenance to preventive maintenance. The frequency of corrective maintenance actions will decrease as the frequency of preventive maintenance actions increases.

- Preventative maintenance provides economic value added to the organization if it achieves cost reduction for system sustainment and improves system operational availability.
- Preventive maintenance reduces labor, materials, and overhead expenses allocated to prerepair logistics downtime.
- Preventative maintenance minimizes the occurrence maintenance actions performed in the field where the system is located that often result in maintenance-induced substantiated part failures.
- Preventative maintenance minimizes lost opportunity costs by reducing the total mean downtime.
- Preventative maintenance increases the efficiency of operating capital by enabling organizations to better allocate scarce resources.

Preventive Maintenance Logistics Downtime Parameters

The dominant opinion in industry and the literature is that the primary benefit from preventive maintenance on the life-cycle costs of the system is the reduction of logistics downtime. There is no question that many of the prerepair logistics downtime events will be eliminated or significantly reduced in duration, yielding significant cost savings and reduction in total downtime. Using our example, the parameters of the prerepair and postrepair logistics downtime models have been reevaluated as shown in Table 6.14. The mean prerepair logistics downtime has decreased from 5.5 hours to one hour, 18%.

TABLE 6.14

Logistical Downtime Parameters
for Preventive Maintenance

	T_{min}	T_{mode}	T_{max}	Λ_{mean}
Λ_{prerep}	1.00	1.25	1.50	1.25
$\Lambda_{postrep}$	0.60	1.00	1.25	0.95

Typically, the postrepair logistics downtime will not see such a dramatic change as reflected in the reduction from 1.5 hours to 0.95 hours, 63%, in our example.

What is often ignored is the effect that preventive maintenance has on the failure math model. The determination of the preventive maintenance interval changes the failure math model dramatically. The preventive maintenance interval for part Delta, selected as the lowest life-cycle cost, is presented to illustrate this point.

Preventive maintenance is based on determination of when preventive maintenance actions will be performed. The question that is often investigated is how to determine what the preventive maintenance interval should be. An analytical approach is presented that applies an existing concept known as the risk threshold, r. The risk threshold is defined to be the risk an organization is willing to accept that a critical part will not fail on the next mission deployment. It is a concept used in aviation since the inception of the development of reliability-centered maintenance in 1962.

Determination of the risk threshold is a management and engineering effort that combines an understanding of the constraints of the organization's business model and the engineering features of the part. Reliability analysis has developed the failure math model for part Delta and has plotted the hazard function, the instantaneous failure rate, in Figure 6.3.

The location parameter for part Delta was previously calculated and is 612 hours. Management and engineering agree that the organization is willing to accept a risk of one unscheduled failure per month. The organization defines one mission duration for the system to be 24 hours. Assuming

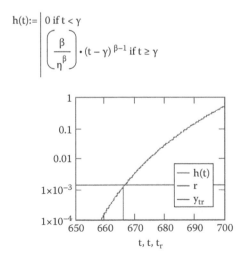

$$h(t):= \begin{vmatrix} 0 \text{ if } t < \gamma \\ \left(\dfrac{\beta}{\eta^\beta}\right) \cdot (t-\gamma)^{\beta-1} \text{ if } t \geq \gamma \end{vmatrix}$$

FIGURE 6.3
Hazard function for part Delta.

Given

r = h(t)

Find(t) → (666.60 ...

tr = 666.60

h(tr) = 0.001389

FIGURE 6.4
Mathcad solve block for part Delta risk-maintenance interval.

a 30 day month, the risk can be expressed as the inverse of the product of 30×24 hours a day, $r = 0.001389$.

The scheduled maintenance interval, t_r, is calculated as the operating hours for which the hazard function is equal to the risk threshold.

$$h(t_r) = r \tag{6.1}$$

The value for the scheduled maintenance interval is found in Mathcad using the routine shown in Figure 6.4.

The hazard function plotted in the previous figure shows the intersection of the risk threshold, the horizontal line, with the hazard function, the ascending line. The scheduled maintenance interval is illustrated by construction, the vertical line from the intersection to the time axis.

The calculated value for the scheduled maintenance interval defines the operating hours that cannot be exceeded in order for the system to continue to operate below the risk threshold. The organization can track the operating hours for a critical part and observe when the part's useful life approaches the risk threshold. This information can be used by the operations and maintenance managers to coordinate scheduling the system for a preventive maintenance action. Maintenance planners can determine from prior experience the amount of time that is required to acquire a spare part and hardware and to schedule the logistical resources that will be required to perform the preventive maintenance action. It is unrealistic to expect that a system will operate up to the precise time designated by the scheduled maintenance interval. An approach to evaluate the estimated period of time during which the system will be removed from service is the uniform probability distribution, ranging from the location parameter of the failure math model to the scheduled maintenance interval. The uniform distribution now defines the new part failure math model. In our example, the parameters of the failure math model will be a uniform distribution for a range of 642 hours to 666 hours.

The implementation of preventive maintenance will also result in a revision to the part repair math model. Recall that a part repair math model based on a maintenance history for a system that experiences downing events that result in the maintenance capability traveling to the system location as well as maintenance actions performed in a controlled environment at a facility is comprised of two distributions. Such part repair math models yield a mean

TABLE 6.15

Part Delta Preventive Maintenance Failure Repair LDT
Reliability Information

	Failure Math Model			Repair Math Model			System		
	T_{min}	T_{max}	γ	η	β	γ	Λ_{prerep}	$\Lambda_{postrep}$	A_O
	642	666	642	0.33	2.26	3.67			
μ			654			3.96	1.250	0.950	0.991

time to repair that is actually a mean of two means. The revised part repair math model will be based on time-to-repair data for the controlled environment at a facility and will differ from the part repair math model developed from all time-to-repair data.

Preventive maintenance reduces the time to perform prerepair and postrepair logistics downtime events to include elimination of logistical downtime events. The preventive maintenance logistics downtime for the system is revised to reflect the benefits of preventive maintenance in Table 6.14.

The part failure, repair, and logistics downtime model parameters for preventive maintenance are provided in Table 6.15 and include the mean time between maintenance actions, formerly the MTBF, the mean time to repair, the mean prerepair and postrepair logistics downtimes, and the operational availability resulting from the implementation of preventive maintenance. Note that the operational availability has increased from 0.9832 to 0.991.

The reliability failure math models for preventive maintenance (f_{pm}) and the time to failure ($f(t)$) are illustrated in Figure 6.5. The lower confidence

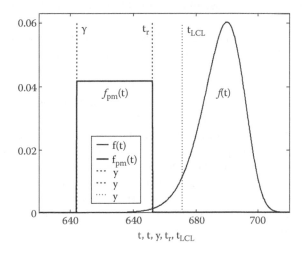

FIGURE 6.5
Part Delta failure math model for preventive maintenance.

limit for time to failure is included in the figure. Preventive maintenance intervals that are based on trend analysis often select the preventive maintenance interval based on the lower confidence limit for time to failure. Notice that the lower confidence limit for time to failure is greater than the scheduled maintenance interval determined by the risk threshold.

Life-Cycle Economic Analysis for Preventive Maintenance

The life-cycle economic analysis preventive maintenance evaluates two options: the baseline and preventive maintenance. The baseline option uses the time-to-failure math model and its point estimate for the MTBF as the duration of the evaluation period. The preventive maintenance option uses the point estimate for the mean time between preventive maintenance actions.

The procedure for developing the part option cost events for preventive maintenance (PM) recognizes a change in the determination of present, uniform recurring, and future amounts. The baseline cost to the organization for the acquisition of a spare part and hardware in the baseline part option cost table are present amounts, as shown in Table 6.16. Preventive maintenance enables an organization to delay the acquisition of spare parts and hardware until they are required; thus those costs become future amounts as materials for repair cost events. The baseline cost to the organization for inventory carrying costs occurred over the duration of the evaluation. Preventive maintenance enables an organization to eliminate or significantly minimize inventory carrying cost. In our example, we assume no cost. Notice also that the preventive maintenance future amounts for pre-repair logistics cost events are significantly lower because the mean prerepair logistics downtime is reduced. Similarly, the future amounts for repair and postrepair logistics cost events are lower for the preventive maintenance option.

The economic analysis for the baseline preventive maintenance options is summarized in Table 6.17. Notice that the future amount for the preventive maintenance option is over twice the future amount for the baseline option. We include the lost opportunity cost in the economic analysis because it is a significant element to the justification for implementing preventive maintenance. The lost opportunity costs for preventive maintenance are half those for the baseline option. It bears noting that preventive maintenance does not eliminate lost opportunity cost. Lost opportunity cost is incurred whenever a system is not available for productive operation. A system is in a downstate when it is in a preventive maintenance action. The equivalent uniform recurring amount demonstrates that the preventive maintenance option for part Delta yields a significantly lower life-cycle cost to the organization.

TABLE 6.16

Part Delta Option Cost Information

	Baseline	PM
P: Present amounts		
Part cost ($)	9,034	
Hardware ($)	837	
A: Uniform recurring amounts		
ICC ($)	2	$0
F: Future amounts		
Prerepair logistics cost events		
Λ_{prerep}	5.53	1.25
Labor ($)	862	195
Cost ($/hr)	156	156
Overhead		
Retrieval vehicle ($)	163	37
Cost ($/hr)	29	29
Repair cost events		
MTTR	4.93	3.96
Labor ($)	3,493	2,808
Cost ($/hr)	709	709
Materials		
Part cost ($)		9,034
Hardware ($)		837
Overhead		
Crane ($)	27	22
Cost ($/hr)	6	6
Facility/bay ($)	58	47
Cost ($/hr)	12	12
Postrepair logistics cost events		
$\Lambda_{postrep}$	1.51	0.95
Labor ($)	343	216
Cost ($/hr)	227	227

TABLE 6.17

Part Delta Economic Analysis

	Baseline[a]	PM[a]
n	688	654
P ($)	−9,871	0
A ($)	−2	0
$PV[A]$ ($)	−1,609	0
F ($)	−4,946	−12,358
L_{Op} ($)	−39,263	−20,215
$PV[F]$ ($)	−43,758	−32,258
NPV ($)	−55,238	−32,258
A_{Equiv} ($)	−81	−50

[a] Economic factors: $r_{IRR} = 12.50\%$; $m = 8{,}400$, and $i_{IRR} = 1.49 \times 10^{-5}$.

Part Life-Cycle Simulation Approach

The part life-cycle simulation approach is presented for two reasons:

1. Reliability simulation software computes the failure, repair, and logistics downtime part math models from the raw data for time to failure, time to repair, and logistics downtime. This provides a significant time savings to the engineer and decreases the opportunity for error in a calculation. It just makes more sense to use a tool that can cut the working time and increase accuracy. Admittedly, the available software programs are relatively expensive. Yet they return their investment by several orders of magnitude on the first reliability investigation.

2. The complexity of assembly math models renders a deterministic approach to be intractable.

Reliability simulation is a progression from part life-cycle simulation and analysis to assembly life-cycle simulation and analysis to system life-cycle simulation analysis. Design applications for reliability simulation include the entire progression and enable the organization to perform trade studies that identify the impact on system reliability parameters based on changes to a critical part in a critical assembly within the system design configuration. Sustainment applications of reliability simulation will often focus on part life-cycle simulation and analysis and may include assembly life-cycle simulation depending on the needs of the organization.

We will illustrate the use of reliability simulation to perform the life-cycle economic analyses previously demonstrated for the deterministic approach.

Several advantages will be realized from the information provided in reliability simulation reports that are not available by deterministic analysis.

There are two variations on reliability simulations: mission duration simulations and life-cycle simulations. Mission duration simulations are run for the duration of the mission or are run to the first part failure. The primary objective of a mission duration simulation is to determine the reliability of the part, assembly, and system. Life-cycle simulations are run for the duration of the useful life of the system. The primary objective of life-cycle simulations is to determine all of the other reliability parameters of a part, assembly, and system, to include availability, mean time between downing events (MTBDE), the MTBF for the part, mean time to repair, mean downtime, number of system downing events, consumption of spare parts, percentage of the time in which the system was operational, degraded mode, or down state, and logistical resource information to include the behavior of sparing strategies, labor constraints, and costs.

This book uses life-cycle simulations only.

The development of simulation math models is presented in Chapter 9. As stated before, this chapter will start with the information provided by the reliability simulation.

Simulation Design

Simulation represents a virtual experiment that follows the same rules for sampling in design of experiments for physical tests. The number of simulation trials must conform to statistical significance. Simulation results are evaluated at confidence level, typically 95%, ranging between 80% and 99%. The examples provided in this chapter use the 95% confidence level. The level of significance, Alpha, is one minus the confidence. The statistically significant number of simulation trials is found by the following equation.

$$n = \frac{\ln(\alpha)}{\ln(R)} \tag{6.2}$$

where R is the reliability allocation for the part.

In our example, however, the reliability requirement was stated in MTBF. One of the few useful applications of the exponential probability distribution is providing an estimate for the value of R. Recall the reliability function for the exponential probability distribution.

$$R = e^{-\lambda \tau} \tag{6.3}$$

$R = 0.96079$ for $\lambda = 1/\text{MTBF} = 0.00333$, and mission duration, $\tau = 12$ hours.

The statistically significant number of simulation trials, n, is 74.9, which we round up to 75. The accepted convention is to always round up, for example

TABLE 6.18

Reliability Life-Cycle Simulation
Design Parameters[a]

UL (yr)	7
f_U/d (hr)	12
d/wk (d)	5
wk/yr (wk)	50
UL (hr)	21,000
θ (hr)	300
λ (fph)	0.00333
τ (hr)	12
R	$0.96079\ e^{-\lambda\tau}$
R	0.96079
α	0.05
n ($\ln(\alpha)/\ln(R)$)	74.9
	75

[a] Requirements: MTBF = 300 hr; $A_O = 0.96$, and lowest cost.

74.1 would also yield 75 simulation trials. The reason is that 74 simulation trials would be below the statistically significant number of simulation trials required.

The duration of a life-cycle simulation trial is equal to the useful life of the system. In our example, the useful life of the system is estimated to be seven years based on a utilization rate of 12 hours per day, five days per week, 50 weeks per year, or 21,000 hours. These calculations are summarized in Table 6.18.

Part Alpha

The deterministic parameters of the part failure, repair, prerepair logistics, and postrepair logistics math models are input to the simulation software program via dialog boxes. A sample of a simulation output report is presented for part Alpha in Table 6.19. Simulation output report formats vary between the commercially available software programs. Only the pertinent information is provided here for the purposes of the part selection life-cycle economic analysis.

TABLE 6.19

Part Alpha Reliability Simulation
Input Information

Failure Math Model			Repair Math Model		
η	β	γ	η	β	γ
35.2	6.67	390	0.56	2.25	3.15

The first simulation report for part Alpha is designated as the baseline simulation. The baseline simulation does not include any provisions for preventive maintenance, spare parts strategies, or other resource constraints. In fact, the default spare parts strategy is that an infinite number of spare parts is available when the part fails. This of course is an unrealistic expectation.

The parameters of interest for the part selection life-cycle economic analysis include the following: availability, MTBDE, mean maintenance time (MMT), MDT, LDT, and the number of part failures. The minimum, mean, and maximum values for the parameters of interest are provided by the simulation routine. We can draw the following observations from the output report for part Alpha:

- Out of 75 life-cycle trial runs at least one life-cycle experienced an operational availability of 0.97043, which is the minimum; the mean operational availability is 0.97560; and at least one life-cycle experienced an operational availability of 0.97685, which is the maximum.

- The mean time between failure ranged from 426.28 hours to 427.37 hours, with a mean of 426.82 hours (The output report uses the term MTBDE regardless of the design configuration. MTBDE is appropriate for assembly design configurations and higher).

- The mean time to repair ranged from 3.58 hours to 3.72 hours with a mean of 3.65 hours (the output report uses the term MMT, for mean maintenance time, regardless of the design configuration. MMT is appropriate for assembly design configurations and higher).

- The mean downtime ranged from 10.13 hours to 11.22 hours, with a mean of 10.68 hours.

- The mean logistical downtime ranged from 6.54 hours to 7.50 hours with a mean of 7.03 hours. The simulation program allowed the input of prerepair logistics downtime and postrepair logistics downtime math models but did not report the distinction between the two events in the report. Instead the report sums the two metrics.

- The mean number of part failures range from 48 to 48 with a mean of 48.

It must be understood that each minimum value is likely to have occurred on different trial runs. For example, one cannot calculate the minimum availability in the report as being equal to the minimum MTBF divided by the sum of the minimum MTBF plus the minimum mean downtime. The same is true for the maximum values. However, the mean availability is equal to the mean MTBF divided by the sum of the mean MTBF and the mean downtime.

The output report includes the standard deviation (SD) and the standard error of the mean (SEM) for each parameter. The standard error of the mean can be used to perform tests of hypotheses for means between two or more simulation reports because of the central limit theorem. The 95% confidence

TABLE 6.20

Part Alpha Reliability Baseline Simulation

	Min	Mean	Max	SD	SEM	95% Confidence Limit	
						CL_S	$CL\mu$
Availability	0.97436	0.97560	0.97685	0.00052	0.00007	0.97473	0.97548
MTBDE	426.28	426.82	427.37	0.23	0.03	426.44	426.77
MMT	3.58	3.65	3.72	0.04	0.005	3.71	3.66
MDT	10.13	10.68	11.22	0.23	0.03	11.06	10.73
LDT	6.54	7.03	7.50			7.35	7.07
Part failures	48	48.00	48	0	0		

limit for the mean, calculated in the last column to the right, is a valid statistical inference.

The same cannot be said for the use of the standard deviation to calculate the 95% confidence limit for the sample because availability is not normally distributed. The same is true for the other parameters. Unfortunately there is no other way provided by the simulation output report to calculate confidence limits for the sample parameters. With great reservation, the 95% confidence limits for the sample parameters are calculated as the only option currently available (Tables 6.20).

Distinctions between the deterministic approach and the simulation approach to describe and evaluate the reliability parameters for each part are illustrated in Figures 6.6–6.8.

The part Alpha operational availability plot shows the deterministic point estimate of the mean and the lower confidence limit. The baseline simulation describes how the operational availability ranges between the minimum and

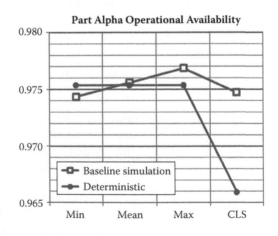

FIGURE 6.6
Part Alpha operational availability comparison.

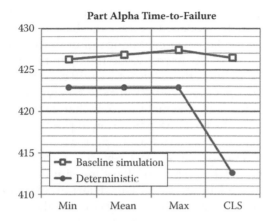

FIGURE 6.7
Part Alpha time-to-failure comparison.

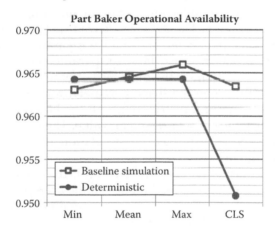

FIGURE 6.8
Part Alpha total downtime comparison.

maximum realized values and the rough estimation for the lower confidence limit. The key issue is the development of an understanding of the measure of dispersion for operational availability. Based on the parameters of the failure math model, the repair math model, and the prerepair and postrepair logistical downtime math models, we can infer that the operational availability for part Alpha will vary over the useful life of the asset. It is reassuring to observe that the lower confidence limit of the sample, though suspect, appears to fit within the range of the minimum and maximum value, which is the expected outcome. The reason that the operational availability for part Alpha will vary is that the occurrences of part failure will vary in accordance with the parameters of the failure math model, and the duration of total downtime will vary as the time to repair, prerepair logistics downtime, and postrepair logistics downtime varies in accordance with the parameters of their respective math models. Contrast the baseline simulation plot to the deterministic plot. The lower confidence limit of the sample represents an abstract expectation of the lowest value the operational availability will achieve over the useful life of the asset. Yet we will expect the preponderance of the values for the operational availability to be clustered about the measures of central tendency for the math models that describe the operational availability. Therefore the baseline simulation appears to present truth that cannot be inferred from the deterministic approach.

The part Alpha time-to-failure plots reveal a curious distinction from the behavior of the deterministic and baseline simulations that was seen in the operational availability plots. The deterministic point estimate of the mean operational availability was nearly equal to the baseline simulation mean operational availability. The time-to-failure plots do not intersect. Indeed the minimum baseline simulation time to failure is greater than the deterministic point estimate of the mean. There is a reason for this observation, which is based on an understanding of the meaning of the parameters of the Weibull distribution. The data for skewed distributions will tend to cluster about the scale parameter, and the scale parameter will be distinct from the mean of the data. Because the shape parameter for part Alpha is relatively high, the expectation for the incidence of time to failure is weighted above the mean. Only when the shape parameter for the Weibull distribution describes a symmetrical distribution will the baseline simulation and deterministic plots intersect near the mean value. The operational availability plots intersected because the baseline simulation deterministic means are composites of several means.

The total downtime plot for part Alpha behaves like the operational availability plot (Figure 6.8). It is the combination of the prerepair logistics downtime, repair, and postrepair logistics downtime math models that define the characterization of the total downtime parameter. Indeed, it is the influence of the total downtime metric on the operational availability that results in the intersection that was observed.

The baseline simulation reports and parameter plots are provided for the other part candidates in the following tables (Tables 6.21–6.28) and figures (Figures 6.9–6.20) without comment. However the information that is contained in the baseline simulation will be applied to the life-cycle economic analysis that follows.

Part Baker

The part Baker reliability simulation input information is shown in Table 6.21. The part Baker reliability baseline simulation is shown in Table 6.22. The part Baker operational availability comparison is shown in Figure 6.9. The part Baker time-to-failure comparison is shown in Figure 6.10. The part Baker total downtime comparison is shown in Figure 6.11.

TABLE 6.21

Part Baker Reliability Simulation
Input Information

Failure Math Model			Repair Math Model		
η	β	γ	η	β	γ
16.8	2.97	295	0.91	2.46	3.65

TABLE 6.22

Part Baker Reliability Baseline Simulation

						95% Confidence Limit	
	Min	Mean	Max	SD	SEM	CL$_s$	CLμ
Availability	0.96304	0.96452	0.96591	0.00067	0.00009	0.96341	0.96437
MTBDE	311.14	311.61	312.06	0.22	0.03	311.26	311.57
MMT	4.37	4.46	4.55	0.04	0.01	4.52	4.47
MDT	11.01	11.46	11.94	0.22	0.03	11.82	11.51
LDT	6.65	7.01	7.39			7.30	7.04
Part failures	65	65.00	65	0	0		

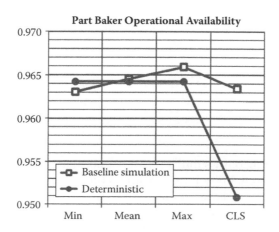

FIGURE 6.9
Part Baker operational availability comparison.

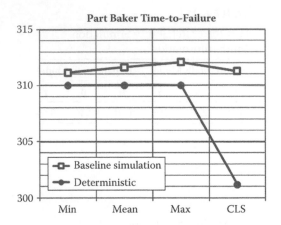

FIGURE 6.10
Part Baker time-to-failure comparison.

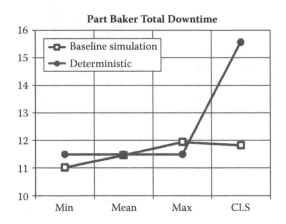

FIGURE 6.11
Part Baker total downtime comparison.

Part Charley

The part Charley reliability simulation input information is shown in Table 6.23. The part Charley reliability baseline simulation is shown in Table 6.24. The part Charley operational availability comparison is shown in Figure 6.12. The part Charley time-to-failure comparison is shown in Figure 6.13. The part Charley total downtime comparison is shown in Figure 6.14.

TABLE 6.23

Part Charley Reliability Simulation
Input Information

Failure Math Model			Repair Math Model		
η	β	γ	η	β	γ
25.3	6.67	450	0.55	2.25	6.25

TABLE 6.24

Part Charley Reliability Baseline Simulation

	Min	Mean	Max	SD	SEM	95% Confidence Limit	
						CL_S	CLμ
Availability	0.97074	0.97190	0.97327	0.00048	0.00006	0.97110	0.97179
MTBDE	474.08	475.03	486.08	2.09	0.27	471.55	474.58
MMT	6.67	6.74	6.82	0.03	0.00	6.80	6.75
MDT	13.05	13.74	14.29	0.24	0.03	14.13	13.79
LDT	6.38	6.99	7.48			7.33	7.04
Part failures	42	42.97	43	0.18	0.02		

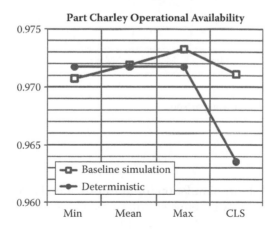

FIGURE 6.12
Part Charley operational availability comparison.

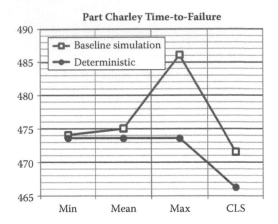

FIGURE 6.13
Part Charley time-to-failure comparison.

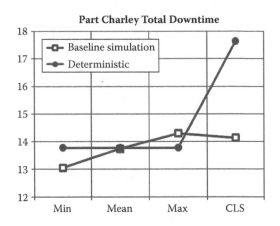

FIGURE 6.14
Part Charley total downtime comparison.

Part Delta

The part Delta reliability simulation input information is shown in Table 6.25. The part Delta reliability baseline simulation is shown in Table 6.26. The part Delta operational availability comparison is shown in Figure 6.15. The part Delta time-to-failure comparison is shown in Figure 6.16. The part Delta total downtime comparison is shown in Figure 6.17.

TABLE 6.25

Part Delta Reliability Simulation
Input Information

Failure Math Model			Repair Math Model		
η	β	γ	η	β	γ
48.9	7.93	642	0.67	3.05	4.33

TABLE 6.26

Part Delta Reliability Baseline Simulation

						95% Confidence Limit	
	Min	Mean	Max	SD	SEM	CL$_S$	CLμ
Availability	0.98226	0.98321	0.98417	0.00040	0.00005	0.98254	0.98312
MTBDE	687.58	699.11	712.64	12.00	1.56	679.11	696.50
MMT	4.85	4.93	5.02	0.04	0.01	5.00	4.94
MDT	11.08	11.94	12.71	0.30	0.04	12.45	12.01
LDT	6.23	7.01	7.69			7.44	7.06
Part failures	29	29.54	30	0.50	0.07		

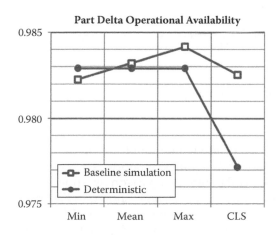

FIGURE 6.15
Part Delta operational availability comparison.

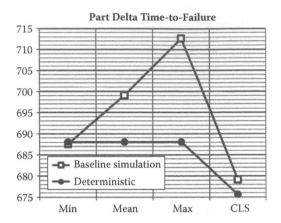

FIGURE 6.16
Part Delta time-to-failure comparison.

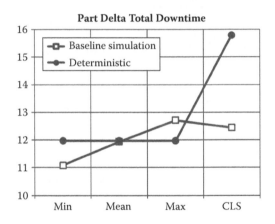

FIGURE 6.17
Part Delta total downtime comparison.

Part Echo

The part Echo reliability simulation input information is shown in Table 6.27. The part Echo reliability baseline simulation is shown in Table 6.28. The part Echo operational availability comparison is shown in Figure 6.18. The part Echo time-to-failure comparison is shown in Figure 6.19. The part Echo total downtime comparison is shown in Figure 6.20.

A comparison of the five parts is provided in the plots in Figures 6.21–6.23 for the operational availability, MTBF, and mean downtime. In the spirit of the expression that a picture is worth a thousand words, the plots shown in Figures 6.21–6.23 better describe the distinctions of the metrics for the five parts.

TABLE 6.27

Part Echo Reliability Simulation
Input Information

Failure Math Model			Repair Math Model		
η	β	γ	η	β	γ
18.5	2.15	401	1.55	2.57	3.67

TABLE 6.28

Part Echo Reliability Baseline Simulation

	Min	Mean	Max	SD	SEM	95% Confidence Limit	
						CL$_S$	CLμ
Availability	0.97103	0.97223	0.97367	0.00061	0.00008	0.97122	0.97209
MTBDE	416.28	422.85	425.98	4.06	0.53	416.08	421.97
MMT	4.85	5.06	5.21	0.08	0.01	5.20	5.08
MDT	11.46	12.08	12.68	0.25	0.03	12.49	12.13
LDT	6.61	7.02	7.46			7.30	7.05
Part failures	48	48.29	49	0.46	0.06		

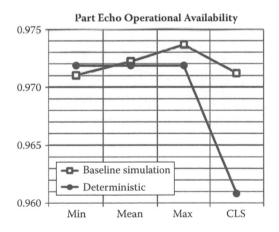

FIGURE 6.18
Part Echo operational availability comparison.

The operational availability plot illustrates the concept of the test of hypothesis for the mean, although not rigorously. For example, in a test of hypothesis for means for two variables, we are essentially evaluating whether the statistically significant range of the confidence limit for the means includes the two means and accept the null hypothesis that the means are statistically significantly the same, or exclude the mean, and reject the

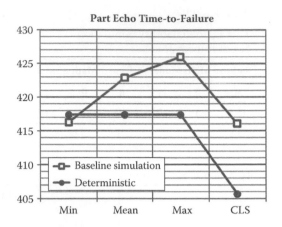

FIGURE 6.19
Part Echo time-to-failure comparison.

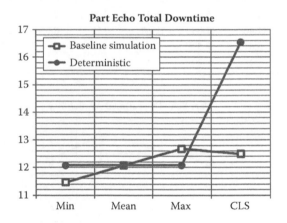

FIGURE 6.20
Part Echo total downtime comparison.

null hypothesis and accept the alternate hypothesis that the means are not statistically significantly the same. When we reject the null hypothesis, we may rank order the parameters. When we are confronted with three or more variables, a test of hypothesis for means normally employs an analysis of variance. But to do that requires the raw data, which is absent from the baseline simulation. All we have are the minimum, mean, and maximum values from 75 life-cycle trial runs. However, we can observe in the following plot that the ranges of the operational availability for parts Alpha, Baker, and Delta do not overlap any other part, and that the ranges of the operational availability for parts Charley and Echo appear to include their respective means. We can draw an inference that the operational availability for parts Charley and Echo are probably equal at some level of confidence. But clearly

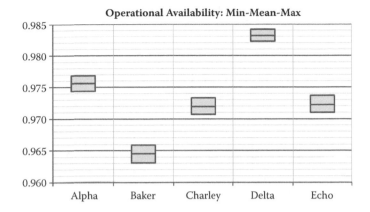

FIGURE 6.21
All parts operational availability comparison.

part Delta provides the highest rank-ordered operational availability and part Baker provides the lowest (Figure 6.21).

A similar approach to comparing the MTBF for the five parts indicates that parts Baker, Charlie, and Delta are significantly different from the others, but that parts Alpha and Echo may be the same. Again we ranked part Delta as having the highest MTBF, and part Baker the lowest (Figure 6.22).

The mean downtime for the five parts does not yield such stark distinctions. It appears that part Charley is significantly different from the other four parts and has the highest mean downtime for a smaller-is-best metric. Parts Baker, Delta, and Echo appear to be similar, with part Alpha half a bubble off. Without statistical rigor, one may surmise that the mean down-

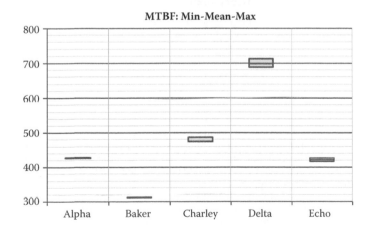

FIGURE 6.22
All parts MTBF comparison.

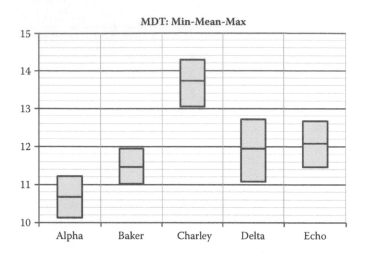

FIGURE 6.23
All parts MDT comparison.

time for parts Alpha, Baker, Delta, and Echo are close enough to each other for practical inference and cannot be rank ordered (Figure 6.23).

Part Life-Cycle Economic Analysis

The life-cycle economic analysis is provided in Table 6.29. Note that the organization's internal rate of return is expressed as 10% APR, with 3,000 compounding periods per year as calculated in the simulation design, yielding an interest rate per compounding period of 0.0000333% per hour.[*]

TABLE 6.29

Economic Information

r_{IRR} (APR, %)	10
m_{IRR} (CP/yr)	3,000
i_{IRR} (r/CP)	3.33×10^{-5}

---

[*] Note to reader: the more observant among you may have noticed in the simulation design that the mission duration and operating scenarios for the five parts are different than for the deterministic life-cycle economic analysis. For those of you that missed that, the economic factors presented on this page would probably have alerted you that the same values are not used from the deterministic approach. This is intentional. It would have been boring to present the simulation approach with all other information identical, so that you would already know the outcome before the economic analysis was performed. With the changes made, you may proceed on with an expectation of a different outcome and maintain your interest in the outcome.

TABLE 6.30

Part Alpha Economic Analysis for Baseline Simulation

	Min ($)	Mean ($)	Max ($)
n_{TTF} (TTF)	426.28	426.82	427.37
A_{ICC} (ICC/hr)		-2.35	
P_{Spare} (burdened part cost (unit spare strategy))		-8,623	
F_{prerep} (Λ_{prerep} hr × (part prerepair LDT labor cost/hr))	-931	-1,021	-1,109
F_{TTR} (TTR × part repair labor cost/hr)	-3,070	-3,070	-3,070
$F_{postrep}$ ($\Lambda_{postrep}$ hr × (part postrepair LDT labor cost/hr))	-407	-407	-407
F_{RV} (Λ_{prerep} hr × (retrieval vehicle cost/hr))	193	193	193
F_{OC} (TTR × (overhead crane cost/hr))	-14	-14	-14
F_{Fac} (TTR × (facility cost/bay hr))	-43	-43	-43
ΣF_i (sum cost for future events)	-4,272	-4,362	-4,450
$PV(A_{ICC})$ ($PV(i_{IIR}$, TTF, $A_{ICC})$)	-995	-996	-997
$PV(\Sigma F_i)$ ($PV(i_{IIR}$, TTF, $\Sigma F_i)$)	-4,212	-4,300	-4,387
NPV(Alpha) (sum of P_{Spare}, $P_V(A_{ICC})$, $P_V(\Sigma F_i)$)	-13,829	-13,919	-14,007
A_{Equiv} (Alpha) ($PMT(i_{IIR}$, 0, $P_V(T_{TFA}))$	-32.67	-32.84	-33.01

Based on the additional information provided by the baseline simulation, the part economic analysis will be expanded to include the minimum, mean, and maximum values for the reliability parameters: MTBF, mean prerepair logistics downtime, MTTR, and mean postrepair logistics downtime. The part Alpha economic analysis for the baseline simulation is provided in detail in Table 6.30.

- n_{TTF} is the minimum, mean, and maximum time to failure.
- A_{ICC} is the uniform recurring amount, which is the inventory carrying cost (ICC) per hour.
- P_{Spare} is the present amount, which is the burden part cost.
- F_{prerep} is the future amount for the prerepair logistics downtime, which is equal to the product of respective values for the minimum, mean, and maximum prerepair logistics downtime hours and the labor cost per hour.
- F_{TTR} is the future amount for the time to repair, which is the product of the respective values for the minimum, mean, and maximum time to repair and the labor cost per hour.
- $F_{postrep}$ is the future amount for the postrepair logistics downtime, which is the product of respective values for the minimum, mean, and maximum postrepair logistics downtime hours and the labor cost per hour.

- F_{RV} is the future amount for the recovery vehicle, which is the product of respective values for the minimum, mean, and maximum prerepair logistics downtime hours and the recovery vehicle cost per hour.

- F_{OC} and F_{Fac} are the future amounts for the overhead crane and facility bay, which is the product of the respective values for the minimum, mean, and maximum time to repair and the overhead crane and facility bay cost per hour.

- $PV(A_{ICC})$ is the present value of the inventory carrying cost, the recurring uniform amount, evaluated for the interest rate per compounding period, number of compounding periods (minimum, mean, and maximum MTBF).

- $PV(\Sigma F_i)$ is the present value of the sum of the future amounts evaluated for the interest rate per compounding period, number of compounding periods (minimum, mean, and maximum MTBF).

- NPV is equal to the sum of the present amount, the present value of the inventory carrying cost, and the present value of the sum of the future amounts.

- A_{Equiv} is the equivalent uniform recurring amount for the NPV evaluated for the interest rate per compounding period, number of compounding periods (minimum, mean, and maximum MTBF), and ranges from \$32.67/hr to \$33.01/hr with a mean of \$32.84/hr.

The summary of the baseline simulation life-cycle economic analysis for the five parts is presented in Table 6.31. Note that the equivalent uniform recurring cost for part Delta peaks at the mean and decreases for the maximum. This is because the magnitude of the maximum MTBF is 712.64 hours.

Once again, the optimum economic life-cycle cost is part Delta, but part Alpha has taken last place.

One more comment about the simulation versus deterministic approach—the baseline simulation took three days from acquisition of the reliability data and economic information, while the deterministic approach took 14 days. The reasons include the following:

- The simulation approach fits the reliability models from the data, the deterministic approach required fitting the data to the reliability math models using the median ranked regression in Excel and completing the calculations in MathCad—the longest pole in the tent.

- The simulation approach performed the cost event calculations that were imported to an Excel worksheet that calculated the NPV and equivalent uniform recurring amounts; the deterministic approach required performing all of the calculations in Excel.

TABLE 6.31

All Part Options Economic Analysis for Baseline Simulation

	Part Alpha			Part Baker			Part Charley			Part Delta			Part Echo		
	Min ($)	Mean ($)	Max ($)	Min ($)	Mean ($)	Max ($)	Min ($)	Mean ($)	Max ($)	Min ($)	Mean ($)	Max ($)	Min ($)	Mean ($)	Max ($)
n_{TTF}	426.28	426.82	427.37	311.14	311.61	312.06	474.08	475.03	486.08	687.58	699.11	712.64	416.28	422.85	425.98
A_{ICC}		−2.35			−2.35			−2.35			−2.35			−2.35	
P_{Spare}		−8,623			−3,803			−5,398			−6,919			−4,376	
F_{prerep}	−931	−1,021	−1,109	−940	−1,018	−1,100	−906	−1,016	0	−881	−1,018	−1,142	−909	−1,020	−1,137
F_{TTR}	−3,070	−3,070	−3,070	−3,674	−3,749	−3,825	−5,613	−5,672	−5,732	−4,078	−4,150	−4,222	−4,078	−4,257	−4,383
$F_{postrep}$	−407	−407	−407	−374	−406	−438	−361	−405	−441	−351	−406	−455	−362	−406	−453
F_{RV}	193	193	193	178	193	208	171	192	209	167	193	216	172	193	215
F_{OC}	−14	−14	−14	−17	−18	−18	−26	−27	−27	−19	−19	−20	−19	−20	−21
F_{Fac}	−43	−43	−43	−52	−53	−54	−79	−80	−80	−57	−58	−59	−57	−60	−61
ΣF_i	−4,272	−4,362	−4,450	−4,879	−5,050	−5,227	−6,814	−7,007	−6,071	−5,220	−5,459	−5,682	−5,253	−5,570	−5,840
$PV(A_{ICC})$	−995	−996	−997	−727	−728	−730	−1,105	−1,108	−1,133	−1,597	−1,624	−1,655	−987	−987	−1,001
$PV(\Sigma F_i)$	−4,212	−4,300	−4,387	−4,828	−4,998	−5,173	−6,708	−6,896	−5,973	−5,101	−5,333	−5,549	−5,179	−5,492	−5,758
NPV	−13,829	−13,919	−14,007	−9,359	−9,529	−9,705	−13,211	−13,402	−12,504	−13,618	−13,875	−14,122	−10,542	−10,854	−11,134
A_{equiv}	−32.67	−32.84	−33.01	−30.24	−30.74	−31.26	−28.09	−28.44	−25.93	−20.03	−20.08	−20.05	−25.50	−25.85	−26.32

System Reliability Analysis

System reliability analysis can only be done by simulation. Consider the following system reliability block diagram comprised of six assemblies (Figure 6.24):

- Assembly 1 is comprised of three parts in a serial design configuration.
- Assembly 2 is comprised of part 2 in a triple active redundant design configuration.
- Assembly 3 is comprised of two serial paths in an active redundant design configuration where each path is comprised of three parts in a serial design configuration, serial in parallel (SiP).
- Assembly 4 is comprised of three parts, each part is in an active redundant design configuration, and each active redundant design configuration is in a serial design configuration, parallel in serial (PiS).

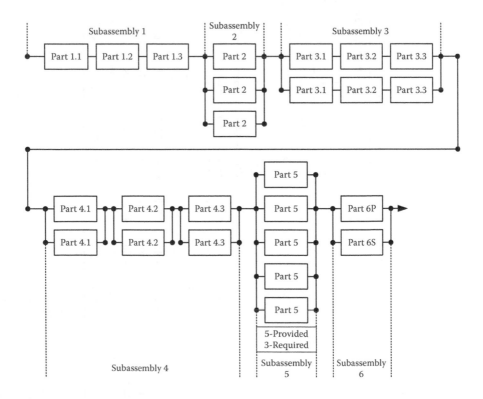

FIGURE 6.24
System reliability block diagram.

- Assembly 5 is comprised of part 5 in a 5 provided, 3 required redundant design configuration, n provided, r required.
- Assembly 6 is a standby design configuration comprised of a primary part and a standby part with a perfect switch.

The parameters of the part failure and repair math models are presented in Table 6.32.

Parameters of the prerepair logistics and postrepair logistics downtime models are provided in Table 6.33.

Before you continue, consider how you would calculate the operational availability of the system, or the MTBDE, or the mean time between maintenance (MTBM). First, let's differentiate between MTBDE and MTBM. MTBDE occurs when one or more assemblies are in a down state. Assembly 1 is straightforward. A serial design configuration enters a down state when one part fails. But the other assemblies follow different rules.

TABLE 6.32

System Baseline Simulation Reliability Information: Part Reliability Information

Assembly	Part	Part Failure Model			Part Repair Model		
		η	β	γ	η	β	γ
Assembly 1	Part 1.1	19.83	7.740	323	1.02	2.33	4.09
	Part 1.2	24.44	5.950	417	0.95	2.51	4.90
	Part 1.3	21.22	6.670	431	0.67	2.26	3.43
Assembly 2	Part 2	18.92	2.830	95	1.08	2.52	4.89
Assembly 3	Part 3.1	25.40	6.180	198	1.24	2.57	6.52
	Part 3.2	37.20	7.030	167	0.71	2.55	5.74
	Part 3.3	45.13	4.920	159	1.14	1.95	7.12
Assembly 4	Part 4.1	36.64	2.700	143	0.83	2.35	3.77
	Part 4.2	34.30	2.320	122	1.19	2.26	5.90
	Part 4.3	29.6	3.070	108	0.89	1.9	6.3
Assembly 5	Part 5	26.24	5.290	126	1.24	2.53	7.42
Assembly 6	Part 6P	42.22	7.054	151	1.14	2.34	5.53
	Part 6S	14.30	2.363	26	1.15	2.25	3.51

TABLE 6.33

System Logistics Downtime Parameters

LDT	T_{min}	T_{mode}	T_{max}
Prerepair	2.90	5.00	9.80
Postrepair	0.50	1.00	2.50

- Assembly 2 enters a down state when all three parts fail during a mission.
- Assembly 3 enters a down state when both paths enter a down state.
- Assembly 4 enters a down state when one parallel pair enters a down state, i.e., both parts fail during a mission.
- Assembly 5 enters a down state when $n - r + 1$ parts fail during a mission ($5 - 3 + 1 = 3$ parts fail).
- Assembly 6 enters a down state when the primary and the standby parts fail during a mission.

The deterministic approach to solving for MTBDE is to develop the system survival function and then solve for the indefinite integral of the system survival function. Try that in Mathcad. I will tell you what happens. Mathcad times out for this system after 20 minutes. But it is worse than that. The survival function applies only when all parts begin the analysis in a new, unaged state, when the system is new. After the first part fails and is replaced in the system, regardless of the assembly in which it is located, the deterministic survival function is no longer valid. At some point in the useful life of the system every part will have different ages.

The deterministic approach cannot distinguish between an unscheduled downing event during scheduled operations and maintenance performed on a system upon completion of a mission to repair parts that failed in the redundant design configurations. The demand for maintenance between scheduled operations is the MTBM. The baseline simulation assumes that failed parts that do not cause a system downing event are replaced at the conclusion of a mission. Maintenance actions will be performed before the system can begin its next mission. The deterministic approach calculates the system operational availability that does not, indeed cannot, survive the first part failure and replacement, just as for the deterministic survival function.

Simulation behaves like a system in use. It calculates time to failure for each part and applies the rules that determine the distinction between a downing event and a maintenance event, and it calculates the prerepair logistics downtime, repair time, and postrepair logistics downtime for each part failure. Simulation cycles the ebb and flow of system behavior over its estimated useful life.

The life-cycle baseline simulation output report for the system is provided in Table 6.34. We observe that the operational availability ranges from 0.80249 to 0.86074 with a mean of 0.83275. Simulation breaks down operational availability into color codes:

1. Green means the system is fully functional with no part failures.
2. Yellow means the system is in a degraded mode with one or more part failures but not a downing event.
3. Red means the system entered a down state.

TABLE 6.34

System Baseline Simulation Results

	Min	Mean	Max	SD	SEM
Availability	0.80249	0.83275	0.86074	0.01275	0.00147
MTBDE	39.61	49.13	59.07	4.71	0.54
MTBM	6.16	6.39	6.63	0.10	0.01
MMT	6.83	6.86	6.88	0.01	0.00
MDT	9.13	9.80	10.56	0.31	0.04
System failures	306	358.77	428	30.21	3.49
Green percentage	13.55	16.02	17.77	0.78	0.09
Yellow percentage	62.95	67.25	71.01	1.76	0.20
Red percentage	13.93	16.72	19.75	1.28	0.15

The sum of the mean color codes is one. The sum of the green and yellow color codes is equal to the mean operational availability. The red color code equals the unavailability, $1 - A_O$.

The MTBDE shows that the assembly enters a down state in the range of 39.61–59.07 operating hours with a mean of 49.13 hours.

The MTBM is not a helpful metric and needs improvement for all reliability simulation programs (Table 6.34). Unless the mission duration is less than six hours the MTBF range from 6.16 to 6.63 operating hours with a mean of 6.39 hours would suggest that maintenance actions are performed on the system while it is operating. Actually, the magnitude of the MTBM is due to multiple part failures each mission. This is confirmed by the discrepancy between the logistical downtime from the deterministic math model and the LDT on the report. The minimum total logistics downtime is 3.4 hours, but the difference between MDT and MMT is 2.94 hours. Assume that the mission duration is 12 hours. At the end of the mission, two parts have failed. The simulation will show that as an MTBF of 12/2 = 6 hours. Meanwhile the LDT for the maintenance event is evenly applied to the two parts where 5.88 hours become 2.94 hours. Fortunately, the simulation does get the MDT right. Notice that the LDT is not included in the simulation output report to avoid confusion. Actually, the three metrics that are valid and useful to life-cycle economic analysis are availability, MTBDE, and MDT.

Assume that the system operational availability is less than the allocated requirement or the sustainment goal. How can the engineer determine the assembly that limits the system operational availability? Simulation provides a weak link analysis that identifies the operational availability for each exit node for each assembly design configuration, as shown in Figure 6.25. The exit node for Assembly 1 has the lowest operational availability. Unless that availability is increased, the system availability cannot improve.

Design options for improving operational availability include changing the part and changing the design configuration to a redundancy. Sustainment

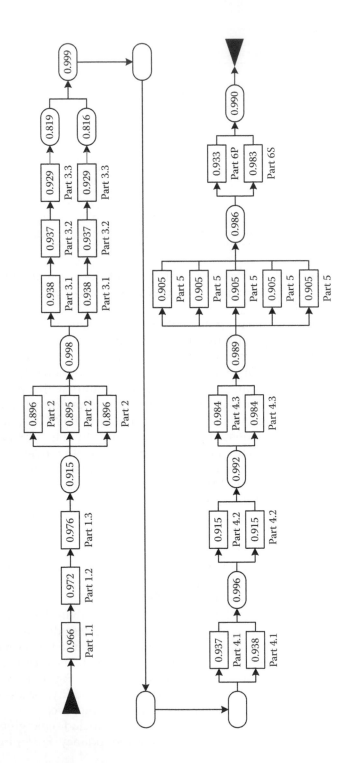

FIGURE 6.25
System baseline simulation weak link analysis.

options include changing the part through a third party vendor and implementation of preventive maintenance. All options achieve the same goal— reduce total downtime.

Preventive Maintenance Approach

The initial Assembly 1 life-cycle simulation confirms the weak link assessment for operational availability, as shown in Table 6.35A. The MTBDE and MDT are 134.32 hours and 12.46 hours, respectively.

The organization defines the risk threshold and calculates the scheduled maintenance interval. The part failure math model is fit to a uniform distribution using the failure math model location parameter, γ, as the minimum time and the scheduled maintenance interval, t_n, as the maximum time.

The repair and logistics downtime math models are fit to the preventive maintenance parameters. Parameters for the reliability math model are presented in Table 6.36.

Assembly 1 baseline life-cycle simulation is run, and the output report is provided in Table 6.37A. The operational availability improves from 0.915 to a range of 0.95349–0.95538 with a mean of 0.95431. The MDT was reduced by half to 6.27 hours.

The plot in Figure 6.26 illustrates the improvement in operational availability resulting from preventive maintenance.

The plot in Figure 6.27 illustrates the improvement in MDT resulting from preventive maintenance.

TABLE 6.35A

Assembly 1 Baseline Simulation Results

	Min	Mean	Max	SD	SEM
Availability	0.91264	0.91507	0.91767	0.00114	0.00013
MTBDE	130.38	134.32	137.60	1.86	0.21
MTBM	127.77	128.11	128.47	0.16	0.02
MMT	4.85	4.94	5.01	0.03	0.00
MDT	12.12	12.46	12.83	0.16	0.02
System failures	140	143.09	147	1.88	0.22

TABLE 6.35B

Spare Parts Utilization

Part	Min	Mean	Max
Part 1.1	59	59	59
Part 1.2	46	46	46
Part 1.3	45	45	45

TABLE 6.36

Assembly 1 Reliability Information with Preventative Maintenance Simulation Information

Part	Part Failure Model		Part Repair Model			System			
Part	γ	t_r	η	β	γ	LDT	T_{min}	T_{mode}	T_{max}
Part 1.1	323	333	0.92	2.45	3.68	Prerep	0.75	1.00	1.33
Part 1.2	417	427	0.86	2.64	4.41	Postrep	0.50	0.66	1.00
Part 1.3	431	440	0.60	2.37	3.09				

TABLE 6.37A

Assembly 1 Simulation Results with Preventive Maintenance

	Min	Mean	Max	SD	SEM
Availability	0.95349	0.95431	0.95538	0.00042	0.00005
MTBDE	128.36	130.94	133.75	1.57	0.18
MTBM	126.73	127.41	127.79	0.39	0.05
MMT	4.37	4.44	4.50	0.03	0.00
MDT	6.16	6.27	6.40	0.06	0.01
System failures	150	153.07	156	1.79	0.21

TABLE 6.37B

Spares Utilization

Part	Min	Mean	Max
Part 1.1	62	62	62
Part 1.2	48	48.28	49
Part 1.3	47	47	47

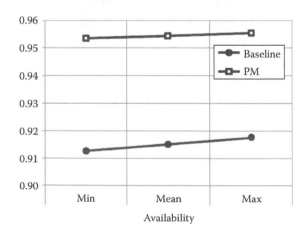

FIGURE 6.26
Simulation availability comparison.

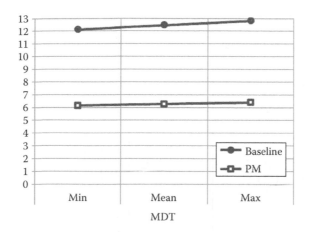

FIGURE 6.27
Simulation MDT comparison.

TABLE 6.38

System Preventive Maintenance Simulation

	Min	Mean	Max	SD	SEM
Availability	0.84170	0.87013	0.89945	0.01184	0.00137
MTBDE	40.08	49.14	61.55	4.39	0.51
MTBM	6.44	6.66	6.88	0.09	0.01
MMT	6.81	6.82	6.86	0.01	0.00
MDT	6.60	7.28	7.79	0.28	0.03
System failures	304	374.35	441	28.31	3.27
Green percentage	15.51	16.77	18.59	0.77	0.09
Yellow percentage	65.58	70.25	74.43	1.77	0.20
Red percentage	10.06	12.99	15.83	1.18	0.14

The system baseline life-cycle simulation is updated to include the failure, repair, and logistics math models for Assembly 1. The output report is presented in Table 6.38. The mean operational availability has increased from 0.83275 to 0.87013.

The weak link analysis for the revised assembly baseline life-cycle simulation shows the improvement in Assembly 1 (Figure 6.28) but also shows that it is still the limiting cause for the system availability.

One of the most powerful benefits of reliability simulation is the ability to evaluate multiple options in a few days rather than perform time-consuming and expensive tests.

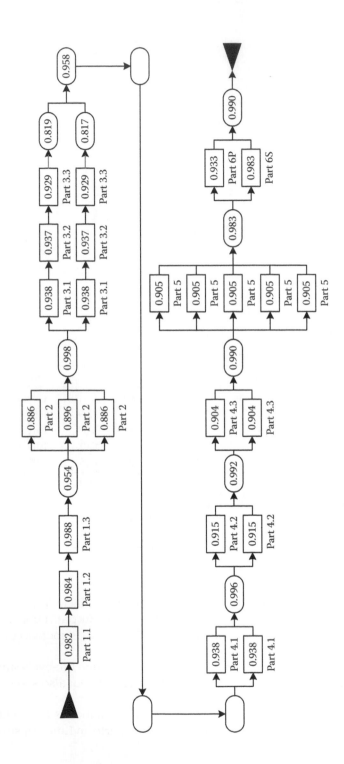

FIGURE 6.28
System preventive maintenance weak link analysis.

Spare Parts Strategy

The spare parts strategy defines the initial inventory for a part, the trigger that initiates the order of new spare parts, the order quantity, and the turn-around time (TAT; Figure 6.29). Assembly 3 has two each of parts 3.1, 3.2, and 3.3.

A spare parts strategy for part 3.1 specifies that one part 3.1 in inventory will be used to replace the first part 3.1 that fails. The trigger occurs when the inventory drops to 0 and one part 3.1 is ordered. The vendor is expected to deliver the part in 15 days, or 360 hours. Simulation will be used to evaluate the baseline strategy and three spare part strategies for parts 3.1, 3.2, and 3.3, as shown in Table 6.39. The baseline strategy is the default setting in reliability simulation software. Although infinite spares are an unrealistic strategy, it does calculate the optimum reliability functions. No spare part strategy can provide better reliability functions.

It is important to acknowledge that deterministic reliability math models do not address the impact on reliability functions caused by sparing strategies.

The baseline life-cycle simulation for Assembly 3 is summarized in Table 6.40 and sets the limits for availability and MDT. MDT in simulation includes time waiting for a spare part when none is in inventory.

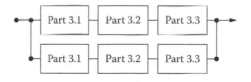

FIGURE 6.29
Assembly 3 - serial-in-parallel design configuration.

TABLE 6.39

Sparing Strategy Options

Baseline Strategy	Infinite Spares		Order	TAT (hr)
	Inventory	Trigger		
1	1	0	1	360
2	2	1	1	360
3	3	1	2	360

TABLE 6.40

Assembly 3 Simulation Infinite Spares

	Min	Mean	Max
Availability	0.93547	0.95660	0.97619
MDT	6.96	8.13	9.35

TABLE 6.41A

Assembly 3 Simulation Sparing Strategy 1:
Availability and MDT

	Min	Mean	Max
Availability	0.00915	0.10852	0.51251
MDT	170.62	11,916	20,808

TABLE 6.41B

Assembly 3 Simulation Sparing Strategy 1: Parts

	Initial No. Spares	New Spares	Min Stock	Avg Stock	Max Stock	End Stock	No. Delays
Part 3.1	1	57	0	0.01	1	0	59
Part 3.2	1	57	0	0	1	0	59
Part 3.3	1	57	0	0	1	0	59

Strategy 1 shocks the system. The mean operational availability is a measly 0.10852, and the MDT is 11,916 hours, roughly half the system's useful life. The number of new spares, 57, is small because so much time is spent waiting for replacements that the system barely has time to operate and allow the parts to fail. The number of delays, 59, is also small, but their durations are huge (Table 6.41A,B).

Strategy 2 shows improvement. The mean operational availability increases to 0.51151, and MDT decreases to 170.99 hours. The number of new spares increases to 114 as the system is able to operate more hours allowing parts to fail. The number of delays increases to 116, but they are shorter in duration (Table 6.42A,B).

TABLE 6.42A

Assembly 3 Simulation Sparing Strategy 2:
Availability and MDT

	Min	Mean	Max
Availability	0.46966	0.51151	0.54659
MDT	158.69	170.99	185.62

TABLE 6.42B

Assembly 3 Simulation Sparing Strategy 2: Parts

	Init No. Spares	New Spares	Min Stock	Avg Stock	Max Stock	End Stock	No. Delays
Part 3.1	2	114	0	0.02	2	0	116
Part 3.2	2	114	0	0.02	2	0	116
Part 3.3	2	114	0	0.02	2	0	116

TABLE 6.43A

Assembly 3 Simulation Sparing Strategy 3: Availability and MDT

	Min	Mean	Max
Availability	0.72715	0.88406	0.99644
MDT	7.75	21.69	39.81

TABLE 6.43B

Assembly 3 Simulation Sparing Strategy 3: Parts

	Init No. Spares	New Spares	Min Stock	Avg Stock	Max Stock	End Stock	No. Delays
Part 3.1	3	170	0	0.05	3	0	119
Part 3.2	3	171	0	0.04	3	0	161
Part 3.3	3	171	0	0.04	3	0	161

Strategy 3 increases the operational availability to 0.884, close to an acceptable magnitude. MDT decreases to 21.69 hours. MDT also shows large variation, ranging from 7.75 to 39.81, indicating that the spare parts strategy is not predictable (Table 6.43A).

Continued analysis can be performed to find the optimum spare parts strategy. Spare part counts are not the only controlled factor. Reducing the turnaround time is another option.

The following plots (Figures 6.30–6.31) for operational availability and total downtime illustrate the impact of the spare part strategies.

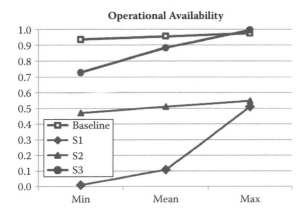

FIGURE 6.30
Assembly 3 sparing strategies operational availability.

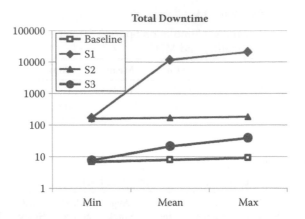

FIGURE 6.31
Assembly 3 sparing strategies total downtime.

Standby Design Configuration

Simulation can calculate reliability parameters for standby design configurations that are not adequately expressed by deterministic approaches (Table 6.44). Assembly 6 is comprised of parts that fit Weibull failure math models and cannot be evaluated by existing equations (Figure 6.32).

The simulation output report finds that the operational availability ranges from 0.98815 to 0.99360 with a mean of 0.99039, as shown in Table 6.45.

The simulation output report also finds the life-cycle consumption rates for the primary and standby parts, as shown in Table 6.46.

TABLE 6.44

Assembly 6 Reliability Information: Baseline

	Standby Assembly	Failure Model		
		η	β	γ
Assembly 6	Part 6P	42.22	7.054	151
	Part 6S	14.30	2.363	26

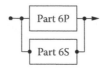

FIGURE 6.32
Assembly 6 standby design configuration.

TABLE 6.45

Assembly 6 Simulation Results: Baseline

	Min	Mean	Max	SD	SEM
Availability	0.98815	0.99039	0.99360	0.00094	0.00011
MTBDE	629.98	677.39	717.83	21.45	2.48
MTBM	153.99	156.37	158.87	1.24	0.14
MMT	5.98	6.08	6.19	0.04	0.00
MDT	4.34	6.57	8.29	0.68	0.08
System failures	29	30.73	33	0.96	0.11
Green percentage	92.23	92.55	92.87	0.14	0.02
Yellow percentage	6.04	6.49	7.13	0.20	0.02
Red percentage	0.64	0.96	1.18	0.09	0.01

TABLE 6.46

Assembly 6 Life-Cycle Parts Consumption

Part	Min	Mean	Max
Part 6P	101	102.27	103
Part 6S	29	30.72	33

TABLE 6.47

Assembly 6 Reliability Information for Exponential Failure Math Model

	Standby Assembly	Failure Model			Exp Equiv	
		η	β	γ	θ	λ
Assembly 6	Part 6P	42.22	7.054	151	190.51	0.00525
	Part 6S	14.30	2.363	26	38.67	0.02586

The time-to-failure data can be fit to the exponential failure math model side by side with the Weibull best fit failure math model, as shown in Table 6.47.

The deterministic exponential survival function for a standby design configuration comprised of unequal parts and a perfect switch is expressed in Mathcad (Figure 6.33). The MTBF, $\theta_{exp} = 229.146$, is found by the indefinite integral of the survival function. The operational availability is calculated to be 0.97213. The exponential deterministic function understates the operational availability by 0.01826 (0.99039–0.97213) and understates the MTBDE by 438.24 hours (677.39–229.15).

$$s_{exp}(t) := e^{-\lambda_p \cdot t} + \left(\frac{\lambda_p}{\lambda_s - \lambda_p}\right) \cdot \left(e^{-\lambda_p \cdot t} - e^{-\lambda_s \cdot t}\right)$$

$$\theta_{exp} := \int_0^\infty S_{exp}(t)dt = 229.146$$

$$A_O := \frac{\theta_{exp}}{\theta_{exp} + MDT} = 0.97213$$

FIGURE 6.33
Assembly 6 mathcad solution for MTBF AO.

Exponential Part's Life-Cycle Consumption

If the error for reliability functions is not bad enough, look at the life-cycle parts consumption determined by the exponential deterministic approach. Part consumption rate is calculated as the product of the part failure rate and the system useful life (UL). The part consumption rates for the primary and standby parts are estimated as follows:

- Primary part: $\lambda_p UL = (0.005525)(21,000) = 110.23$
- Standby part: $\lambda_s UL = (0.02586)(21,000) = 543.01$

The error between the baseline life-cycle simulations and the exponential primary part life-cycle consumption for the primary part is only 8 (110.23 − 102.27), but the error for the standby part is atrocious at 512 (543.01 − 30.72), an overstatement of 1,752%!

7

Reliability-Centered Maintenance

Reliability-centered maintenance (RCM) picks up with the part failure mode. Chapters 2 and 3 cover in detail how to perform a criticality analysis, develop a critical items list, and identify failure mechanism sources, failure mechanisms, failure modes, and failure effects. RCM focuses on failure modes and how to measure them. The RCM road map shown in Figure 7.1 illustrates the process that is required before implementing a reliability-centered maintenance solution. See the reliability analysis in Chapters 2 and 3 for additional details on what must be done leading up to implementing a RCM maintenance solution.

Choosing the Correct Maintenance Plan

The optimal maintenance plan is based upon how the part can be monitored. There are three distinct maintenance plans that can be implemented:

- Condition-based maintenance (CBM)
- Time-directed maintenance (TDM)
- No-maintenance solution (NoM)

For each plan there are certain criteria that must be met in order to implement. When a condition indicator can be placed upon a part and there is one failure mode, then condition-based maintenance (CBM) is the ideal maintenance solution for the part. If a condition indicator cannot be used, then proceed to see whether time-directed maintenance will work for the part. TDM can only be used if there is one failure mode caused by one failure mechanism. These are the maintenance plans that can be applied to a part, but they are not the only action that can take place. Another alternative is run to failure (RTF). This action is chosen when the cost of implementing a RCM solution is greater than the cost to the organization for the part failing. If running the part to failure is not an option, then administrative actions need to take place to change the design to prevent said part failure. This is known as the no-maintenance solution (NoM). Figure 7.2 illustrates the thought process behind choosing the optimal maintenance solution for the part.

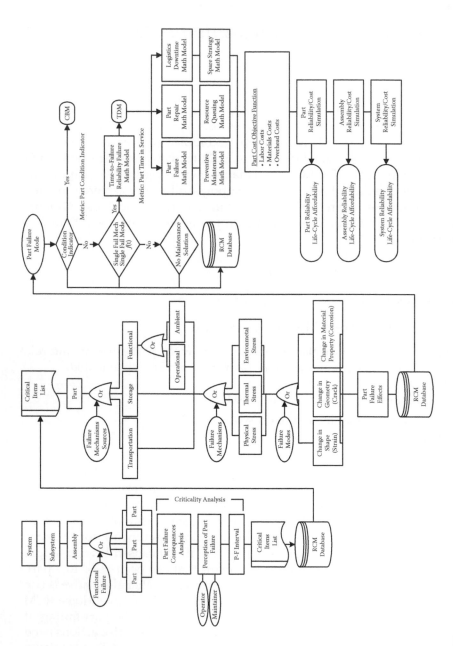

FIGURE 7.1
RCM roadmap.

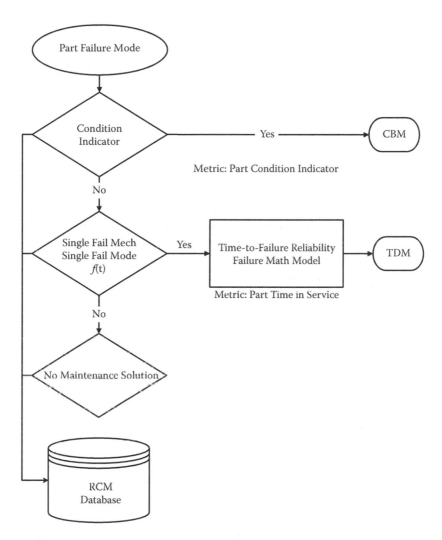

FIGURE 7.2
RCM decision criteria.

Condition-Based Maintenance

Condition-based maintenance uses a condition indicator to measure a failure mode of a part.

- What must be done?

  A condition indicator must be installed to detect/monitor a failure mode that occurs when a failure mechanism acts upon a part. This failure mode is determined through reliability analysis. The condition indicator only monitors one failure mode.

- Why must it be done?

  The condition indicator detects a failure mode that is forming on a part and notifies maintenance that an action must take place before the system goes into a down state due to part failure. This is being proactive trying to minimize the amount of downtime for the system.

- How must it be done?

  The time between perception of failure and occurrence of the failed state (P-F interval) is used to determine when a part has consumed its useful life and is unable to perform the next mission successfully. When the consumed life of the part is used to the point that the system is unable to complete next mission, the part must be replaced, otherwise the system will go into a down state and will be unable to complete the task.

- Who must do the work?

  The maintenance personnel and engineers will do the work. The data from the condition indicator are used to determine whether the part is able to complete the next mission. The engineers and maintenance personnel work together to perform an analysis that determines at which point on the P-F interval that maintenance action must take place. When the threshold is reached, the engineer works with the maintenance personnel to schedule a maintenance action before the part fails. Maintenance will then perform the required maintenance action, restoring the system to full functionality.

- When must it be done?

  The maintenance personnel will perform the maintenance when scheduled. This is based upon the P-F interval in that the perception of failure has exceeded the desired threshold for the part and that the next mission will cause the part to go from the perception of a potential issue to part failure. Ideally, one would like to give maintenance advance notice to properly schedule the maintenance task to minimize the amount of downtime associated with the maintenance action without allowing the system to run until the part fails and maintenance is scrambling around trying to find all the right tools and parts.

Time-Directed Maintenance

Time-directed maintenance is the simplest of the RCM plans to implement, since it is based upon time or cycle intervals, but there are certain conditions that must be met to base maintenance strictly upon an interval.

- What must be done?

  Determine the optimal interval after which the part must be replaced. This interval can be based upon time or a set number of cycles that the system experiences. This maintenance philosophy can be implemented when there is only one failure mode and one failure mechanism.

- Why must it be done?

  Parts continually wear as the system is used. Some parts have one failure mode and one failure mechanism, which are based upon a time metric. Time-directed maintenance is based upon a certain number of cycles or time that the system is in use before maintenance actions take place. If the system were to be allowed to operate for longer than the determined amount of time or cycles, the risk of the system failing is higher than the company's risk threshold.

- How must it be done?

  Engineering and maintenance work together to determine the optimal number of cycles that the system can experience before the risk threshold level exceeded the allowable limit for the system. The cycle monitor is installed and notifies the correct personnel when the next maintenance action is approaching so it can be properly scheduled to minimize the amount of downtime.

- Who must do the work?

  Maintenance personnel will do a majority of the work. When the system is approaching the limit of the number of cycles, maintenance will schedule the maintenance task before the maximum number of allowable cycles is exceeded. This allows the action to be scheduled at maintenance's convenience.

- When must it be done?

  Maintenance actions should take place during the predefined interval before the system exceeds its maximum allowable risk. This interval is the time from which a notification is received that the system has used a certain percentage of its useful life to the time that the system will exceed the allowable risk threshold. During that time maintenance will need to schedule and perform the required actions before the acceptable risk level is exceeded to minimize the possibility of part failure.

No-Maintenance Solution

A no-maintenance solution is used when it is not technologically or economically feasible to implement condition-based maintenance or time-directed maintenance. Another term for this solution is run to failure (RTF).

- What must be done?

  With a no-maintenance solution there is nothing that can be done concerning maintenance to detect the failure modes of the part. The only option is to redesign the system.

- Why must it be done?

  It is not feasible to insert a condition indicator on the asset, and time to failure is not an adequate method to replace the part. Another reason for NoM is that the cost of implementing a CBM or TDM solution is greater than the cost of part failure.

- How must it be done?

  Not applicable for this maintenance solution. There is no work that can be done until a part fails and needs to be repaired or replaced.

- Who must do the work?

  Permission for system redesign must come from management. The engineers and maintainers must be part of the system redesign if management decides that is an option. Otherwise there is no work for this maintenance solution until the part fails and maintenance repairs the system.

- When must it be done?

  There is no work for this solution until the part fails. At that time the part needs to be replaced or repaired to restore the system.

RCM Challenges

Implementing RCM has hurdles that need to be overcome to be successfully implemented. The biggest challenge is convincing management to change the maintenance practices. People do not like to change, but changing an existing practice that has been done that way for as long as everyone can remember is going to raise concerns and be met with some resistance. Implementing RCM will stress the organizational structure and existing maintenance practice because it forces a different thought process, in other words, it changes the way things are done. Money is the life blood of an organization, so why would someone want to replace a part when it still has some unused utility left on it? Typically, management is opposed to this because they want to maximize their profit. Removing and replacing a part before it fails seems like a waste of unused utility to management, but what does it cost the organization when the system goes down and you are on a tight schedule with deadlines approaching and you are unable to meet the customer's needs? What level of risk are you willing to accept? Also the initial cost to implement a RCM solution takes over a year to break even. Then there is always the skeptic who thinks statistics equals lies and there is no need to do anything differently.

Management may be the toughest hurdle to overcome, but there is also another challenge that must be faced. The analytical tools for RCM are not readily available or understood. There is no one "right" way to implement RCM, because each part or system is different, and providing a one-solution fits-all will not work. Providing guidelines to determine what solution is best for the part or system is possible because the needs of each organization are different. It is critical to understand what it is the organization is trying to accomplish before a RCM solution will be profitable.

RCM Benefits

Benefits from RCM include:

- Reduces marginal costs of operation leading to increased profit margin
- Reduces risk of employee hazards resulting from part failures
- Conserves resources because tasks cost less than random corrective maintenance activities
- Allows optimization of spare parts inventory
- Reduces the incidence of lost production opportunity caused by random corrective maintenance activities
- Reduces collateral damage caused by random part failures
- Reduces risk of litigation from employee lost time accidents and fatalities

RCM provides intangible benefits to the organization:

- Reduces employee hazards and improves operations employee morale and productivity
- Increases productivity and improves organization's reputation for quality and on-time delivery
- Improves tasks on scheduled basis and improves maintenance employee morale and productivity
- Improves working relationships between maintenance and operations

8

Reliability Database

The objective of reliability databases is to render all analyses simple regardless of the complexity of information. Reliability databases provide the ability to acquire and use large amounts of diverse data that are seamlessly integrated with computational features that generate all of the analyses discussed in this book. Ideally, failure data are automatically acquired from system built-in test equipment; repair and logistical downtime data are accessed from management information systems; and all cost rates are acquired from accounting sources without the need for manual entry from documents.

The ideal reliability database should have the following features:

- Part identification, including its design configuration within its assembly through the system level, which could be imported from design analysis; bill of materials; and design art for design database applications and from vendor-provided documentation with electronic transfer of the content for sustainment applications

- Part criticality analysis, including functional fault analysis from the system through the design hierarchies to the part, which uses input scores to automatically rank parts that qualify for inclusion in the critical items list, in design and historical maintenance analysis for frequency and severity of part failure in sustainment

- Part qualitative failure analysis, including sources of failure mechanisms, failure mechanisms and their metrics, failure modes and their metrics, failure effects and their metrics, and assembly failure effects

- Part quantitative failure analysis, including automated empirical data acquisition for failure experiments directly transferred to the database in design and part failure empirical data acquisition from electronic maintenance records and built-in test equipment in sustainment

The reliability database should have the computational capability to fit part failure data to a failure math model, survival function, reliability function, and hazard function. The database should calculate a part's mean time between failure (MTBF) and the lower confidence limit for time to failure.

- Part qualitative repair analysis, including detailed work instructions for the performance of the part repair, which includes required labor skill categories, tools, working environment for each event, special instructions for calibration or certification, in design for maintainability and sustainment
- Part quantitative repair analysis, including experimental time-to-repair data in design and historical data that distinguish between corrective and preventive maintenance actions in sustainment

The reliability database should have the computational capability to fit part repair data to a repair math model. The database should calculate part mean time to repair and the upper confidence limit for time to repair including distinctions between corrective and preventive maintenance actions.

- System qualitative logistics downtime analysis, including prerepair and postrepair logistics downtime events, which specifies required labor skill categories, fault detection fault isolation methods, and requirements for specialty tools and facilities that are specific to corrective and preventive maintenance
- System quantitative logistics downtime analysis, including data to fit prerepair and postrepair logistics downtime math models that are specific to corrective and preventive maintenance

The reliability database should have the computational capability to fit corrective and preventive maintenance logistics downtime data to prerepair and postrepair logistics downtime math models. The database should calculate the system mean prerepair and postrepair logistics downtime and the respective upper confidence limits.

- Part economic information, including hourly rates for the cost to the organization for labor, materials, and overhead expenses broken down by prerepair logistics downtime, repair time, and postrepair logistics downtime

The reliability database should have the computational capability to convert part economic information into hourly rates for the cost to the organization for labor, materials, and overhead expenses for the performance of prerepair logistics downtime, repair time, and postrepair logistics downtime.

Reliability databases should be compatible with commercially available reliability simulation software programs. Reliability simulation software programs that include all of the elements of the reliability database are the preferred solution, if it can be done. Such an approach would require a global vertical integration of all data acquisition across all systems and all users—an unrealistic expectation. Therefore, reliability databases that can generate text and data fields that are directly imported into commercially available reliability simulation software programs are a practical and realistic expectation.

The applicable ability of commercially available database software programs to reliability simulation and analysis is not well understood. Database development in industry is ad hoc, ranging from spreadsheets and commercially available database software programs to internally designed and developed databases. The immaturity of the reliability discipline explains the lack of reliability databases. The closest attempts at reliability databases in industry are the input files designed by reliability software program developers. Government agencies, and by extension their contractors, have gone a long way toward standardizing the development of databases for failure analysis but remain off target for using those databases to feed reliability simulation math models. The optimum short-term solution for a reliability database is to adapt existing commercially available software programs to the extent that they can be used to import data to commercially available reliability simulation software programs. The proposed reliability database presented in the annotated database flow diagrams (Figures 8.1 and 8.2) serves two objectives:

1. Provide guidance to evaluate the potential for adaptability of commercially available database programs to export reliability data and economic information to commercially available reliability simulation software programs
2. Provide guidance to the development of a reliability database by design and sustainment engineers and managers or by developers of integrated reliability simulation software programs

The proposed reliability database is presented in seven segments that are matched with the logic flow diagrams for the implementation of a reliability-based life-cycle economic analysis program.

Reliability database legend

Segment 1: criticality analysis

Segment 2: qualitative failure analysis

Segment 3: quantitative failure analysis

Segment 4: qualitative and quantitative repair analysis

Segment 5: qualitative and quantitative prerepair logistics downtime analysis

Segment 6: qualitative and quantitative postrepair logistics downtime analysis

Segment 7: spare parts strategy

TABLE 8.1

Reliability Database Legend

Input Text field	
Input Data field	
Logic Process	
Function	
Output Report	
Path	
Output Data	
Connect to	

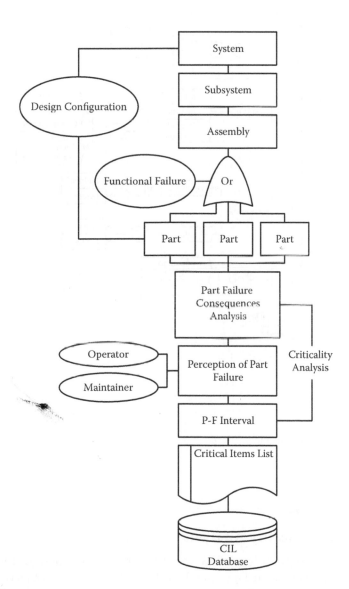

FIGURE 8.1
Criticality analysis.

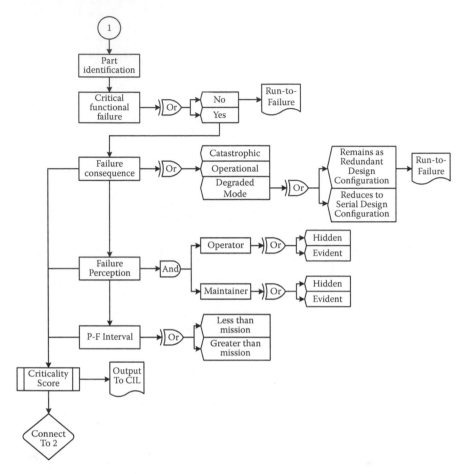

FIGURE 8.2
Reliability database—criticality analysis.

Segment 1: Criticality Analysis

The reliability program logic flow chart for the development of the critical items list is presented in Figures 8.1 and 8.2. The primary objective of the critical items list is to reduce the entire system design hierarchy into a list of critical part candidates. Critical parts are provided with a designation for part identification that includes part nomenclature and its location within the system design hierarchy. Noncritical parts are forwarded to the run-to-failure classification. The criticality analysis evaluates the candidate parts for inclusion in the critical items list.

- Part identification

  Part identification is a process input to the reliability database that should ideally be transferred from the bill of materials or manually input in design or input from part documentation that accompanies the system in sustainment.

- Critical functional failure

  Critical function failure traces the system design hierarchy down to the critical assemblies to identify parts within the critical assembly that will be designated as critical or noncritical.

- Failure consequence, catastrophic, operational, degraded mode, run to failure

  The proposed modified Moubray criticality analysis applied by the evaluation team manually enters the failure consequence.

- Failure perception, hidden, evident by operator and maintainer

  The evaluation team manually enters the corresponding failure perception for the failure consequence.

- Time between perception of failure and occurrence of the failed state (P-F interval), less than mission duration, greater than mission duration

  The evaluation team manually enters the corresponding P-F interval for the failure perception as it pertains to the operator and maintainer.

- Criticality score

  The reliability database assigns the criticality score to the part based on the criticality analysis inputs.

- Critical items list

  The scored part is entered into a reliability database and is automatically rank ordered.

Segment 2: Qualitative Failure Analysis

The reliability program logic flow chart for the performance of the qualitative failure analysis is presented in Figures 8.3 and 8.4. The objectives of the qualitative failure analysis are to identify the sources of failure mechanisms that act on the critical part, the failure mechanisms from each source, the failure modes caused by each failure mechanism, the symptomatic failure effect of each failure mode, and the assembly failure effect. The failure report analysis corrective action (FRACAS) is the proposed approach for the acquisition of this information, root cause failure analysis, and recommended corrective action, which contributes to the failure database.

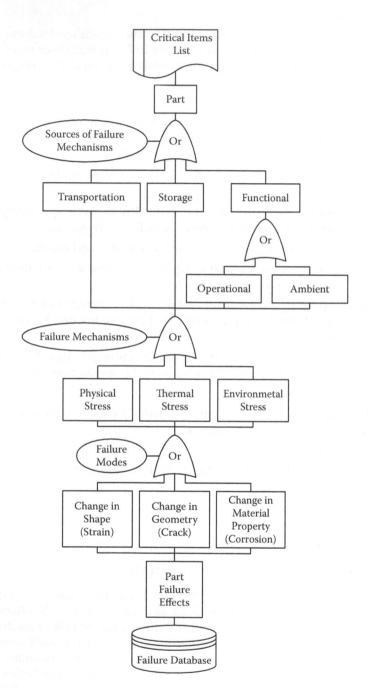

FIGURE 8.3
Qualitative failure analysis.

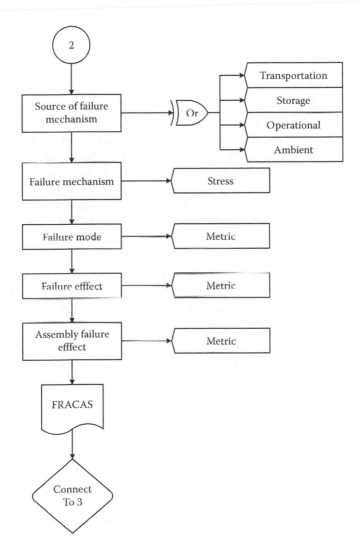

FIGURE 8.4
Reliability database—qualitative failure analysis.

- Source of failure mechanism, transportation, storage, operational, ambient

  All sources of potential failure mechanisms are manually entered by the evaluation team.

- Failure mechanism, stress

  Failure mechanisms that act on the critical part for each source are identified by the evaluation team including the metric that defines the stress load. (Note: "time in use" is not a stress load although time to failure is commonly applied to the development of failure math models.)

- Failure mode, metric

  Failure modes, the damage done to the critical part by the failure mechanism, are identified by the evaluation team, including the metric that defines the damage. Metrics may include continuous variables (dimensions for strain), attribute variables (number of cracks), combination of continuous and attribute variables (number of cracks and their dimensions), and subjective observations (presence of corrosion).

- Failure effect, metric

  Failure effects, the symptoms of each failure mode, are identified by the evaluation team, including how the failure mode degrades the design or capacity of the critical part.

- Assembly failure effect

  The assembly failure effects are identified by the evaluation team, including how the part failure degrades the designer capacity of the assembly.

FRACAS

The information and data for each critical part are stored in a reliability database, including comment fields that present the findings of the evaluation team. These comments address the corrective action based on the analysis of the critical part failure, including: the qualitative repair, prerepair logistics downtime, and postrepair logistics downtime analysis.

The reliability program logic flow chart for the performance of the quantitative failure, repair, and logistics downtime analyses is presented in Figure 8.5. The objectives of the quantitative failure, repair, and logistics downtime analyses are to design the experimental approach to empirically acquire failure, repair, and logistics downtime data and fit failure, repair, and logistics downtime math models. The reliability program logic flow chart is presented for the three parallel paths: failure, repair, and logistics downtime, which can be and often are performed concurrently and by different engineering and management teams.

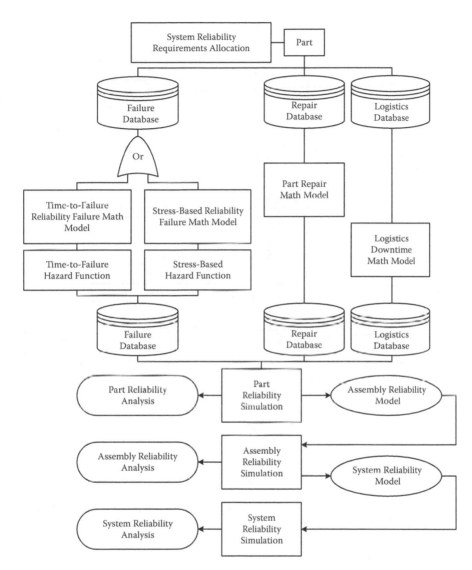

FIGURE 8.5
Quantitative reliability analysis.

Segment 3: Quantitative Failure Analysis

- Reliability allocation, metric

  Reliability allocation resulting from systems engineering allocation of system requirements in design and by determination of the reliability goal in sustainment is entered as a metric, including the mission reliability and lower confidence level, the operational

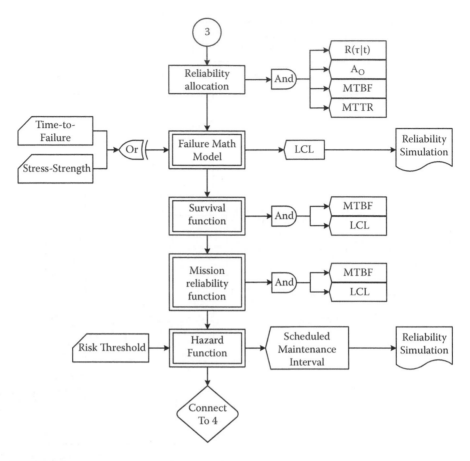

FIGURE 8.6
Reliability database—quantitative failure analysis.

availability and lower confidence level, the mean time between failure and lower confidence level, and the mean time to repair and lower confidence level (Figure 8.6).

- Failure math model

  The reliability database computes the parameters of the failure math model using the input data from empirical investigations for time to failure and stress-strength experiments. The reliability database computes and stores the lower confidence limit for time to failure. The parameters of the failure math model are exportable to the reliability simulation software program.

- Survival function

  The reliability database fits the part survival function from the failure math model. The reliability database computes and stores the MTBF and lower confidence limit.

- Mission reliability function

  The reliability database fits the part mission reliability function from the failure math model. The reliability database computes and stores the point estimate of the mean for mission reliability and lower confidence limit.

- Hazard function

  The reliability database fits the part hazard function from the failure math model. The organization's risk threshold is an input to the reliability database. The reliability database computes and stores the scheduled maintenance interval using the risk threshold as a constraint to the hazard function. The reliability database exports the scheduled maintenance interval in the form of a preventive maintenance failure math model to the reliability simulation software program.

Segment 4: Quantitative Repair Analysis

- Repair events

  The part repair events from the qualitative repair analysis to find the experimental approach to empirically acquire time-to-repair data for corrective and preventive maintenance actions (Figure 8.7).

- Repair math model

  The reliability database imports empirical time-to-repair data and fits the repair math models for corrective and preventive maintenance actions. The reliability database computes the mean time to repair and upper confidence limit for corrective and preventive maintenance actions. The reliability database exports the mean time to repair and upper confidence limit for corrective and preventive maintenance actions to the reliability simulation software program.

- Labor skills category: repair maintenance

  Labor skill categories for the performance of repair maintenance, developed in the qualitative repair analysis, are input as economic analysis information. Labor skills category allocations for repair maintenance are input as economic analysis data.

- Labor skills category: hourly cost repair maintenance

  Labor skills category hourly rates for repair maintenance are input as economic analysis data. The reliability database computes each labor skill category hourly cost as a function of the labor skills category allocation and the labor skills category hourly rate. The reliability database computes the combined hourly cost to perform

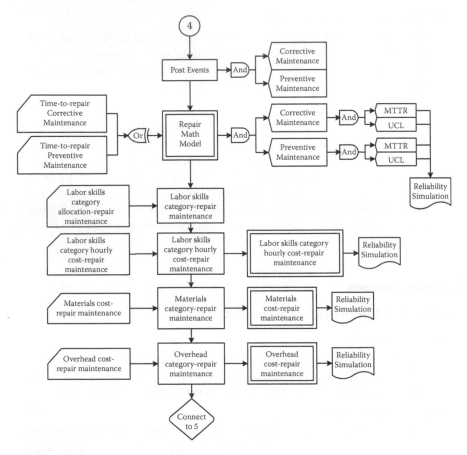

FIGURE 8.7
Reliability database—quantitative repair analysis.

repair maintenance for all of the labor skills categories. The reliability database exports the combined hourly cost to perform repair maintenance for all labor categories to the reliability simulation software program.

- Materials category: repair maintenance, metric

  Materials and their respective costs to the organization required to perform repair maintenance, except spare parts, are input from the qualitative repair analysis. The reliability database calculates the materials cost per repair maintenance event. The reliability database exports the materials cost per repair event to the reliability simulation software program.

- Overhead category: repair maintenance, metric

  Overhead resources and their respective costs to the organization required to perform repair maintenance are input from the qualitative repair analysis. The reliability database calculates the overhead costs per repair maintenance event. The reliability database exports the overhead costs per repair event to the reliability simulation software program.

Segment 5: Prerepair Logistics Downtime Quantitative Analysis

- Prerepair logistics downtime events

  The part prerepair logistics downtime events from the qualitative repair analysis defined the subject matter expert Delphi approach to subjectively acquire minimum, mode, and maximum prerepair logistics downtime data for corrective and preventive maintenance actions (Figure 8.8).

- Prerepair logistics downtime math model

  The reliability database imports empirical prerepair logistics downtime data and fits the prerepair logistics downtime math models for corrective and preventive maintenance actions. The reliability database computes the mean prerepair logistics downtime and upper confidence limit for corrective and preventive maintenance actions. The reliability database exports the mean prerepair logistics downtime and upper confidence limit for corrective and preventive maintenance actions to the reliability simulation software program.

- Labor skills prerepair logistics downtime

  Labor skill categories for the performance of prerepair logistics downtime events, developed in the qualitative logistics downtime analysis, are input as economic analysis information. Labor skills category allocations for prerepair logistics events are input as economic analysis data.

- Labor skills category hourly cost prerepair logistics downtime

  Labor skills category hourly rates for prerepair logistics downtime events are input as economic analysis data. The reliability database computes each labor skill category hourly cost as a function of the labor skills category allocation and the labor skills category hourly

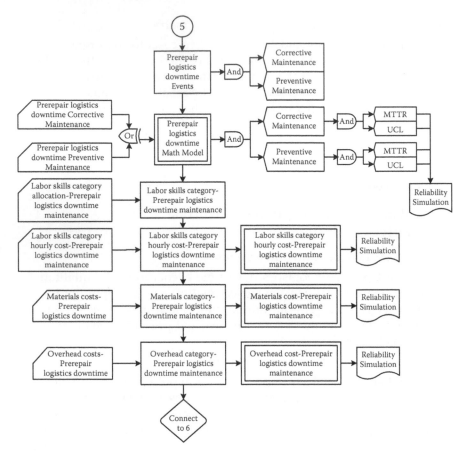

FIGURE 8.8
Reliability database—prerepair logistics downtime quantitative analysis.

rate. The reliability database computes the combined hourly cost to perform prerepair logistics downtime for all of the labor skills categories. The reliability database exports the combined hourly cost to perform prerepair logistics downtime for all labor categories to the reliability simulation software program.

- Materials prerepair logistics downtime

  Materials and their respective costs to the organization required to perform prerepair logistics downtime are input from the qualitative repair analysis. The reliability database calculates the materials cost per prerepair logistics downtime event. The reliability database exports the materials cost per prerepair logistics downtime event to the reliability simulation software program.

- Overhead prerepair logistics downtime

  Overhead resources and their respective costs to the organization required to perform prerepair logistics downtime events are input from the qualitative repair analysis. The reliability database calculates the overhead costs per prerepair logistics downtime event. The reliability database exports the overhead costs per prerepair logistics downtime event to the reliability simulation software program.

Segment 6: Postrepair Logistics Downtime Quantitative Analysis

- Postrepair logistics downtime events

  The part postrepair logistics downtime events from the qualitative repair analysis defined the subject matter expert Delphi approach to subjectively acquire minimum, mode, and maximum postrepair logistics downtime data for corrective and preventive maintenance actions (Figure 8.9).

- Postrepair logistics downtime math model

  The reliability database imports empirical postrepair logistics downtime data and fits the postrepair logistics downtime math models for corrective and preventive maintenance actions. The reliability database computes the mean postrepair logistics downtime and upper confidence limit for corrective and preventive maintenance actions. The reliability database exports the mean postrepair logistics downtime and upper confidence limit for corrective and preventive maintenance actions to the reliability simulation software program.

- Labor skills postrepair logistics downtime

  Labor skill categories for the performance of postrepair logistics downtime events, developed in the qualitative logistics downtime analysis, are input as economic analysis information. Labor skills category allocations for postrepair logistics events are input as economic analysis data.

- Labor skills category hourly cost postrepair logistics downtime

  Labor skills category hourly rates for postrepair logistics downtime events are input as economic analysis data. The reliability database computes each labor skill category hourly cost as a function of the labor skills category allocation and the labor skills category hourly rate. The reliability database computes the combined hourly cost to

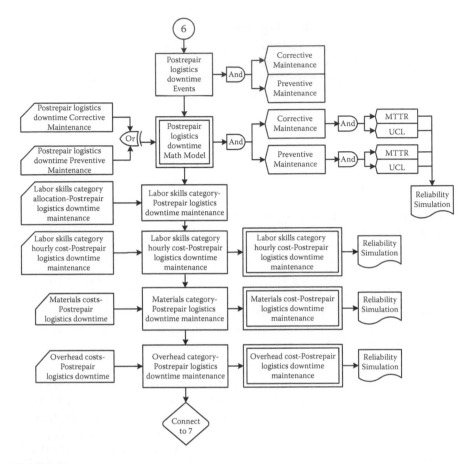

FIGURE 8.9
Reliability database—postrepair logistics downtime quantitative analysis.

perform postrepair logistics downtime for all of the labor skills categories. The reliability database exports the combined hourly cost to perform postrepair logistics downtime for all labor categories to the reliability simulation software program.

- Materials postrepair logistics downtime

  Materials and their respective costs to the organization required to perform postrepair logistics downtime are input from the qualitative repair analysis. The reliability database calculates the materials cost per postrepair logistics downtime event. The reliability database exports the materials cost per postrepair logistics downtime event to the reliability simulation software program.

- Overhead postrepair logistics downtime

  Overhead resources and their respective costs to the organization required to perform postrepair logistics downtime events are input from the qualitative repair analysis. The reliability database calculates the overhead costs per postrepair logistics downtime event. The reliability database exports the overhead costs per postrepair logistics downtime event to the reliability simulation software program.

Segment 7: Spare Part Strategy

Spare parts strategy is separate from materials because of the unique features it provides in the determination of system availability and mean time between downing events (MTBDE). Spare parts represent a significant cost to the organization that must be measured, controlled, and managed, not only for acquisition and storage but also for the impact that a lack of spare parts will have on MTBDE. There are infinite combinations an organization can investigate for initial part stock in inventory, triggers to initiate acquisition of new parts, the quantity of new parts ordered, the unit price negotiated for quantity discounts, inventory carrying costs, cost of an emergency spare part, and turnaround time.

Multiple spare parts strategies are the norm in the complex world of supply chain operations. However each path is essentially the same. It is the linking of the various paths to the end result that provides the complexity and the potential error in the estimates. There are extensive database options that do an excellent job of dealing with the complexity of spare parts economics. The proposed reliability database for spare parts is linear (Figure 8.10). It contains the essential path for the required inputs to the reliability simulation software program for the critical part's replacement strategy.

- Spare parts strategy

  The spare parts strategy is input to the reliability database and exported to the reliability simulation software program. Spare parts strategies can be unit spare parts, economic order quantities, and any variation of the two.

- Spare part unit cost

  Spare part unit cost is entered into the reliability database and exported to the reliability simulation software program. It can be qualified by the emergency unit cost of a spare part. The spare part unit cost and its time of occurrence are input for corrective maintenance and preventive maintenance.

- Spare part initial inventory

  Initial inventory for spare parts is entered into the reliability database and exported to the reliability simulation software program.

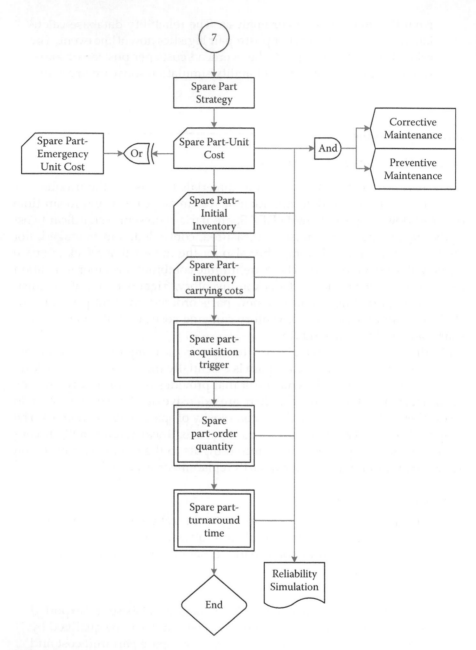

FIGURE 8.10
Reliability database—spare part strategy.

- Spare part acquisition trigger

  The spare part acquisition trigger is entered into the reliability database and exported to the reliability simulation software program.

- Spare part order quantity

  The spare part order quantity is entered into the reliability database and exported to the reliability simulation software program.

- Spare part turnaround time

  The spare part turnaround time is entered into the reliability database and exported to the reliability simulation software program.

9

Reliability Simulation and Analysis

Reliability simulations and analysis software are powerful tools that can aid engineering in decision making. Example simulations for a part, assembly, and system will be shown to illustrate the process that is required to run a reliability simulation and analysis.

Part Reliability Simulations

Part Failure Analysis

The first example is on the part level. Suppose a part has time-to-failure data, shown in Table 9.1.

Inputting the data into reliability software is the first step in performing simulations. Most reliability software can fit data to a three-parameter Weibull distribution. Unlike the long tedious process used to solve the three-parameter Weibull by hand, a best-fit distribution analysis can rank order multiple distributions based upon their correlation coefficient in a matter of seconds. Many different distributions can be examined; a few of them are shown in Table 9.2.

- Three-parameter Weibull
- Two-parameter Weibull
- Normal
- Log-normal
- Rayleigh
- Exponential
- Gamma

The three-parameter Weibull distribution is the most robust and often the highest one or two of the distributions examined. This part is no exception. The three-parameter Weibull has the lowest residual y value and is used as the distribution for the failure model. Open a blank reliability block diagram (RBD) worksheet. On this new worksheet create a new block (part). Each block has many different fields that can be used to input the part description, assembly, cost, and even the part number. The failure math model can be

TABLE 9.1

Part Time-to-Failure Data

TTF (hr)		
250	293	229
262	273	245
288	297	217
289	264	301
267	278	294
282	292	231
229	277	294
258	292	248
248	272	278

TABLE 9.2

Best-Fit Distribution Analysis Results

	Distributions					
	Weibull (three parameters)	Normal	Rayleigh (one parameter)	Weibull (two parameters)	Log-normal	Exponential (one parameter)
β	7.5474		2.0000	13.0517		
η	170.0520		352.3421	278.9038		160.5715
γ	108.4902					
μ		268.4444			5.5886	
σ		24.4591			0.0933	
λ						0.0062
ρ	0.9847	0.9699	0.9843	0.9843	0.9635	−0.8270
ρ^2	0.9696	0.9407	0.9689	0.9689	0.9283	0.6840
Residual Y	1.1637	1.5103	26.6203	1.1903	1.8499	29.4527

linked to the RBD block within most reliability simulation software. Link the failure math model to the block on the RBD. If it is preferable to manually enter the failure math model, choose the desired distribution from the drop down menu and enter the failure parameters. A drop down menu for the failure distributions will typically contain several different options listed below.

- Three-parameter Weibull
- Normal
- Log-normal
- Rayleigh
- Exponential
- Uniform
- Constant time
- Triangular

Part Repair Analysis

The same part has time-to-repair (TTR) data, provided in Table 9.3.

The same procedure performed on the failure data is done to fit the repair data. Results from the best-fit distribution analysis shows the three-parameter Weibull is the best fit for the data. Weibull parameters for the data are provided in Table 9.4.

Unlike the failure model, the results for the repair model must be manually entered into the part's RBD. At this time it is not possible to link repair data to a part. This is primarily due to the nature of the repair model. Repair parameters are manually entered into the maintenance section of the dialog boxes. It could be listed under repair, replace, or restoration time depending on the software or nature of the part. With the repair time option selected, choose the distribution from the drop down menu. In this case it's the Weibull. Enter the values for Beta, Eta, and Gamma into their respective fields. Most simulation tools provide the following distributions for the repair model.

TABLE 9.3

Part Time-to-Repair Data

TTR (hr)			
2.02	2.20	2.10	2.07
2.12	3.05	1.88	2.28
2.32	1.95	2.10	2.02
2.45	2.20	2.23	2.42
2.07	2.23	2.72	1.85
2.60	1.93	2.75	2.17
2.13	2.33	2.45	2.18
1.98	2.68	2.60	3.18
1.93	2.32	2.48	2.87
2.85	1.82	1.90	

TABLE 9.4

Weibull Parameters

Weibull (three parameters)	
β	1.4563
η	0.5664
γ	1.7849
ρ	0.9984
ρ^2	0.9968
Residual Y	0.1813

TABLE 9.5

Mean Logistics Downtime (MLDT)

MLDT (hr)
8.33

- Three-parameter Weibull
- Normal
- Log-normal
- Rayleigh
- Exponential
- Uniform
- Constant time
- Triangular

The last input for the simulation is the logistical down time. Some simulation tools allow for the input of both prerepair and postrepair logistical downtime, others do not. If both are not an option, combine the prerepair and postrepair logistical downtime and input it into the software. The mean logistical downtime for this part is shown in Table 9.5.

To input the logistics downtime data into the software, select the desired distribution and input the corresponding parameters. Typically the logistical downtime is the triangular distribution. Now all the data needed to run a simple part simulation are inserted into the simulation software.

Simulation Results

There are many different reliability and availability calculations that can be selected and performed during a reliability simulation. The most common calculations selected are listed below.

- Reliability
- Availability
- Mean time between failures (MTBF)
- Failure frequency
- Downtime
- Number of failures
- Costs
- Confidence limits

The RBD for the part is shown in Figure 9.1.

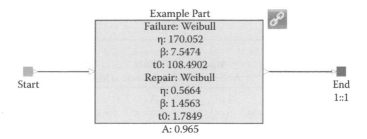

FIGURE 9.1
Part RBD.

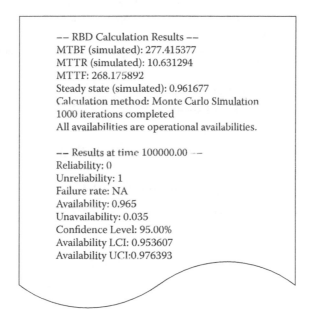

FIGURE 9.2
Part simulation output.

The simulation was run for 100,000 hours with 1,000 iterations. This far exceeds the number of sample runs that are needed to have a statistically significant number of trials. An example output from the simulation is shown in Figure 9.2 and Table 9.6. There are many other reports and outputs that can be formulated to fit various needs.

Most of the analysis for a single part can be performed by hand, but the reliability simulation software makes it much quicker and easier to input the data and have the analysis done in a few hours rather than days. At this point, you would have:

TABLE 9.6

Simulation Output Table

Time (Hr)	Availability	Hazard Rate	Mean Availability	Failure Frequency	Total Downtime	Expected No. of Failures	Capacity
0	1	NA	1	NA	0	0	100
10,000	0.967	3,660.289555	0.962394	3,539.5	376.05503	35.395	96.7
20,000	0.971	3,694.747683	0.962125	3,587.6	757.492643	71.271	97.1
30,000	0.963	3,723.05296	0.962052	3,585.3	1,138.447686	107.124	96.3
40,000	0.96	3,739.0625	0.962002	3,589.5	1,519.938403	143.019	96
50,000	0.969	3,699.380805	0.961981	3,584.7	1,900.962756	178.866	96.9
60,000	0.961	3,732.570239	0.961963	3,587	2,282.22545	214.736	96.1
70,000	0.963	3,724.091381	0.961951	3,586.3	2,663.398085	250.599	96.3
80,000	0.964	3,716.078838	0.961949	3,582.3	3,044.064984	286.422	96.4
90,000	0.971	3,696.292482	0.961937	3,589.1	3,425.70856	322.313	97.1
100,000	0.965	3,714.92228	0.961932	3,584.9	3,806.797132	358.162	96.5

- Developed a failure math model.
- Created a new RBD worksheet.
- Inserted a new part and named it.
- Inserted or linked the failure math model to the part.
- Inserted the repair math model.
- Inserted the logistical downtime math model.
- Inserted any cost information.
- Ran the simulation.
- Generated reports.

All of the data input and simulation steps can be applied to the system level simulations.

Assembly and System Simulations

Reliability modeling for assemblies and systems is where the most gain comes from for the simulation software. It is impossible to perform reliability models for assemblies by hand, so simulation tools are a must.

Consider the RBD in Figure 9.3 for a system.

The system is comprised of four different assemblies and ten unique parts. Each part has its own failure and repair model, and the system has a mean

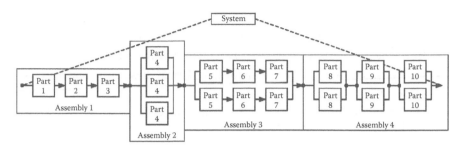

FIGURE 9.3

System RBD.

TABLE 9.7

Part Data for the System

Part	Failure Model			MTBF, θ	Repair Model			Mean TTR, μ	MLDT (hr)
	β	η	γ		β	η	γ		
Part 1	4.22	31.13	29	57.30	2.12	1.59	3.50	4.91	8.33
Part 2	5.17	26.91	32	56.76	1.96	0.96	2.46	3.31	8.33
Part 3	6.33	42.82	21	60.84	2.25	1.63	2.87	4.31	8.33
Part 4	1.96	12.30	36	46.91	2.43	1.65	2.49	3.95	8.33
Part 5	5.43	19.45	42	59.94	2.45	1.72	3.70	5.23	8.33
Part 6	4.79	13.62	38	50.47	2.47	1.81	3.00	4.61	8.33
Part 7	3.94	17.68	48	64.01	1.90	1.02	3.96	4.87	8.33
Part 8	1.78	9.56	19	27.51	2.14	1.48	2.69	4.00	8.33
Part 9	2.37	11.84	26	36.49	2.41	1.62	2.16	3.60	8.33
Part 10	2.09	13.44	20	31.90	1.98	1.33	2.13	3.31	8.33

logistical downtime. Parameters for the parts failure, repair, and logistical downtime are shown in Table 9.7.

The RBD must be constructed within the reliability simulation software. It is easier to build each assembly on its own separate block diagram and then link all of the assemblies at the system level once each assembly has be created. All of the reliability data must be manually inputted into each block in the RBD. It is possible to import all the data directly from a spreadsheet. This requires tedious work setting up the worksheets and labeling the blocks within the simulation tool properly. Otherwise it becomes a gigantic mess and the data are not inserted into the correct data field. Using a database will allow the exportation of templates that will feed data into the correct fields for the simulation software if the parts are not modeled within the software.

The following steps are performed to build a system within a reliability simulation software package.

- Determine the RBD for the system.
- Develop failure and repair math models for each part.
- Build each assembly on its own RBD worksheet.
- Link or insert the failure, repair, and logistical downtime math models to each part.
  - If using a database to import all the data, be careful to link all the block/parts to the right data.
- Insert any cost Information.
- Build the system RBD.
  - Link all the assemblies to the system diagram.
  - Each assembly is a block on the system level RBD.
- Run the simulation.
- Generate reports.

The diagram in Figure 9.4 shows all of the assemblies linked at the system level and the corresponding availability for each assembly.

Spares, maintenance personal, and maintenance tasks can also be inputted into the simulations. In this example they are not. The purpose of this chapter was to show the basics of what goes into a simulation.

The simulation was run for 100,000 hours with 1,000 iterations. Results from the system simulation are shown in Figure 9.5 and Table 9.8.

The simulation can perform this in a matter of minutes. If attempted by hand, it would take someone years to perform the calculations. There are numerous graphs and charts that can be constructed using the tools in the simulation software. A graph of availability versus time for the system and each assembly is shown in Figure 9.6.

The report below provides valuable insight on the availability, downtime, and number of failures for each part and assembly and is an example of some of the analyses that can be performed.

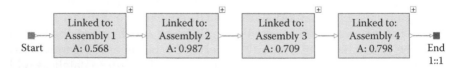

FIGURE 9.4
RBD: system level.

-- RBD Calculation Results --
MTBF (simulated): 45.165153
MTTR (simulated): 17.746321
MTTF: 27.262495
Steady state (simulated): 0.607079
Calculation method: Monte Carlo Simulation
1000 iterations completed
All availabilities are operational availabilities.

-- Results at time 100000.00 --
Reliability: 0
Unreliability: 1
Failure rate: NA
Availability: 0.326
Unavailability: 0.674
Confidence Level: 95.00%
Availability LCI: 0.296941
Availability UCI:0.355059

FIGURE 9.5
System simulation output.

TABLE 9.8

System Output Table

Time (Hr)	Availability	Hazard Rate	Mean Availability	Failure Frequency	Total Downtime	Expected No. of Failures	Capacity
0	1.000	NA	1.0000	NA	0.0000	0.000	100.0
10,000	0.318	148,142.7673	0.3221	47,109.4	6,778.9896	471.094	31.8
20,000	0.327	144,709.1743	0.3220	47,319.9	13,559.8220	944.293	32.7
30,000	0.340	139,221.4706	0.3221	47,335.3	20,336.8651	1,417.646	34.0
40,000	0.286	165,518.8811	0.3220	47,338.4	27,118.2482	1,891.030	28.6
50,000	0.339	139,536.8732	0.3220	47,303.0	33,897.7841	2,364.060	33.9
60,000	0.336	140,744.3452	0.3221	47,290.1	40,676.2128	2,836.961	33.6
70,000	0.332	142,256.3253	0.3221	47,229.1	47,455.1921	3,309.252	33.2
80,000	0.327	144,710.7034	0.3220	47,320.4	54,238.9704	3,782.456	32.7
90,000	0.369	128,180.4878	0.3220	47,298.6	61,017.0971	4,255.442	36.9
100,000	0.326	144,932.2086	0.3221	47,247.9	67,786.1125	4,727.921	32.6

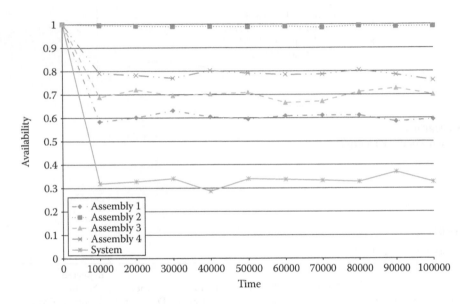

FIGURE 9.6
Availability versus time.

System Simulation Output

Diagram: Assembly 1

Description	Total Downtime	Availability	Number of Failures
Part 1	18761.072533	0.787000	1417.310000
Part 2	17016.076514	0.823000	1461.837000
Part 3	12040.690914	0.888000	1445.514000

Diagram: Assembly 2

Description	Total Downtime	Availability	Number of Failures
Part 4	20748.747892	0.803000	1689.373000
Part 4	20748.551794	0.767000	1689.219000
Part 4	20745.797215	0.783000	1689.072000

Diagram: Assembly 3

Description	Total Downtime	Availability	Number of Failures
Part 5	18438.268119	0.800000	1360.354000
Part 5	18436.470658	0.804000	1360.221000
Part 6	20396.019695	0.808000	1576.816000
Part 6	20396.099636	0.814000	1576.727000
Part 7	30947.741378	0.686000	2510.013000
Part 7	30947.352712	0.699000	2509.965000

Diagram: Assembly 4

Description	Total Downtime	Availability	Number of Failures
Part 10	26726.590338	0.718000	2296.490000
Part 10	26724.525703	0.728000	2296.305000
Part 8	30949.144805	0.662000	2509.941000
Part 8	30950.331781	0.724000	2510.165000
Part 9	24625.031762	0.756000	2064.896000
Part 9	24625.176545	0.767000	2064.851000

Diagram: System

Description	Total Downtime	Availability	Number of Failures
Assembly 1	40687.102597	0.568000	3051.801000
Assembly 2	902.270784	0.987000	219.787000
Assembly 3	30491.615582	0.709000	3605.179000
Assembly 4	21153.317947	0.798000	3254.339000

FIGURE 9.7
System simulation output.

Summary

Reliability simulation software is expensive. However, after the first assembly or system-level project performed with the software, the company will see a return on the investment. Without reliability simulation software tools it would be impossible to run system-level analysis.

Appendix A: Reliability Failure Math Models and Reliability Functions

Two approaches are presented to fit time-to-failure (TTF) data to reliability failure math models: the exponential and the Weibull probability density functions. Time-to-failure data can be either complete or time censored. Complete data describe a sample of part failures with no survivors.

- In design, complete data result from an experiment in which all of the parts are run to failure.
- In sustainability, complete data results from historical records for all parts that have failed.

Time-censored data described a sample of part failures with survivors.

- In design, time-censored data results from an experiment in which insufficient time is available to run all parts to failure.
- In sustainability, time-censored data results from historical records for all parts that have failed and all parts that are currently operating.

Exponential Reliability Math Modeling Approach

The exponential distribution has been applied to failure analysis since the first days of reliability analysis. It is conveniently fit by the arithmetic mean of the part's time to failure. The mean time between failure (MTBF) is considered a key reliability parameter. Digital and electronic reliability analysis uses the exponential distribution exclusively.

Complete Data

Complete data are acquired from empirical investigations (historical and experimental) in which all test articles are run to failure.

Table A.1 provides complete data for time to failure in hours for a part.

The Excel descriptive statistics routine was used to calculate the arithmetic mean of the sample as well as the standard error of the mean (SE), the sample median, the sample mode, the sample standard deviation, the sample variance, the sample kurtosis and skewness, the sample range, the

TABLE A.1

Time-to-Failure Data

TTF (hr)				
1,026	980	963	1,001	975
1,006	966	972	1,002	972
1,007	941	949	1,002	996
1,032	1,024	960	1,007	1,037
999	989	1,006	989	978
1,016	1,004	958	976	953
1,006	974	977	972	991
988	992	1,023	992	987
1,016	996	994	947	1,023

TABLE A.2

Time-to-Failure Descriptive Statistics

Mean	990.3
Standard error	3.531
Median	992
Mode	1,006
Standard deviation	23.69
Sample variance	561
Kurtosis	−0.571
Skewness	−0.114
Range	96
Minimum	941
Maximum	1,037
Sum	44,564
Count	45

sample minimum and maximum time to failure, the sum of the sample data, and the sample size (Table A.2).

The mean is the arithmetic average of the sum of the time to failure divided by the sample size. The standard error of the mean is equal to the sample standard deviation divided by the square root of the sample size. The median time to failure is the 50th percentile. The mode is the most prevalent value for time to failure and often does not exist for continuous data. The standard deviation is the square root of the variance. The variance is equal to the sum of the difference between each time-to-failure data point and the mean time to failure squared. The kurtosis is the standard kurtosis that compares the peak of the data to a normal curve with the calculated values for the mean and standard deviation. When the standard kurtosis is equal to zero, the peak of the sample data is equal to a corresponding normal distribution. A negative standard kurtosis means that the peak of the sample data is flatter than the corresponding normal distribution. A positive standard kurtosis means that the peak of the sample data is higher than the corresponding normal distribution. When the standard skewness is equal to zero, the sample data are symmetrical about the mean, and the mean will equal the median. When the standard skewness is negative, the sample data are a negative skew with a tail that extends toward the origin of the *x*-axis. When the standard skewness is positive, the sample data have a positive skew with a tail that extends toward infinity. The importance of the kurtosis and skewness is its ability to describe the shape of the sample distribution. The time-to-failure data presented here will fit a distribution that is flatter than the normal distribution and will be negatively skewed toward the origin of the *x*-axis.

The exponential failure math model can be fit directly from the descriptive statistics for time to failure. The sample mean is the mean time between failure. The inverse of the mean, λ, is the failure rate, expressed as failures per hour. The lower confidence limit (LCI) of the MTBF for complete data is equal to the MTBF plus the product of the Student's *t* statistic evaluated at the level of significance, α, and the degrees of freedom, $v = n - 1$, and the standard error of the mean. The level of significance is equal to one less the confidence level of the investigation. The MTBF and the lower confidence limit for the complete data time to failure are presented in Table A.3.

The frequency distribution for the time to failure is calculated in Excel using the histogram routine. The histogram routine requires that a "bin" be designated that includes the minimum and maximum range of the histogram and the width of the class intervals. The minimum and maximum range of the histogram should include all of the values of the time to failure. The width of the class intervals is determined by the range of the time-to-failure data. The frequency distribution is shown in Table A.4.

The frequency distribution table was plotted in a histogram (Figure A.1). The negative kurtosis of the time-to-failure data is not intuitively apparent from the histogram, but the negative skew is evident visually.

TABLE A.3

Exponential Time-to-Failure
Parameter and LCL Mean

Math Model	
θ	990.31
λ	0.001010
α	0.05
v	44
$t_{\alpha,v}$	−1.680
θ_{LCL}	984.38
λ_{LCL}	0.001016

TABLE A.4

Time-to-Failure Frequency Distribution

TTF	$f_s(t)$
940	0
950	1
960	1
970	3
980	6
990	7
1,000	10
1,010	11
1,020	3
1,030	3
1,040	0

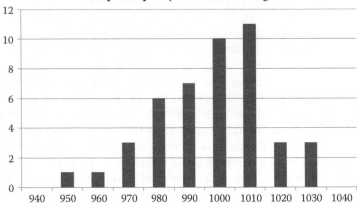

FIGURE A.1
Time-to-failure frequency distribution histogram.

The exponential failure math model for the sample time to failure is expressed in the following equation.

$$f_{exp}(t) = \lambda e^{-\lambda t}$$

$$f_{exp}(t) = 0.00101 e^{-0.00101t} \tag{A.1}$$

The exponential failure math model and the value of the failure rate is plotted in Figure A.2.

The exponential cumulative failure math model is expressed in the following equation.

Given

$\alpha = 1 - e^{-\lambda \cdot t}$

Find(t) → 50.795749431991289909

$t_{LCL} := 50.79575$

$F_{cd}(t_{LCL}) = 0.05$

FIGURE A.2
Exponential failure math model and hazard function.

$$F_{exp}(t) = 1 - e^{-\lambda t}$$

$$F_{exp}(t) = 1 - e^{-0.00101t}$$

(A.2)

The exponential survival function, often called the life function, is the complement of the exponential cumulative failure math model and is expressed in the following equation.

$$S_{exp}(t) = e^{-\lambda t}$$

$$S_{exp}(t) = e^{0.00101t}$$

(A.3)

The lower confidence limit of the time to failure is the expected time to first failure of the part (Figure A.3). It is the time to failure that meets the condition that the cumulative failure math model evaluated for that time will be equal to the level of significance. The goal seek routine in Excel can be used

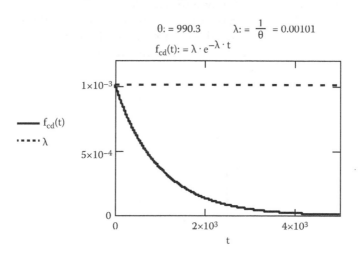

$0 := 990.3$ $\lambda := \dfrac{1}{\theta} = 0.00101$

$f_{cd}(t) := \lambda \cdot e^{-\lambda \cdot t}$

FIGURE A.3
Mathcad solver for time-to-failure LCL.

TABLE A.5

Goal Seek Time-to-Failure
Sample LCL

	Goal Seek
α	0.05
LCL	51.26
$F(LCL)$	0.050

to find the lower confidence limit of the time to failure. An estimate for the lower confidence limit is entered into a cell. The equation for the cumulative failure math model is entered into another cell. Goal seek sets the equation equal to the level of significance by changing the value of the estimate. The goal seek solution is presented in Table A.5.

The standard deviation for the exponential probability distribution is equal to the mean. Note that this is not the case in the descriptive statistics table (Table A.2) [990 ≠ 23.69].

The median time to failure for the exponential probability distribution is equal to the MTBF times the natural logarithm of two. Note that the median calculated below is significantly different from the median in Table A.2.

$$M_{exp} = \theta \ln(2) = (990.31)(0.693) = 686.42 \text{ [rather that 992]} \qquad (A.4)$$

Note that the descriptive statistics provided by Excel assume that the data are normally distributed.

The plots for the exponential cumulative failure math model and the survival functions are provided in Figure A.4. Note that the value of the median occurs at the intersection of the two functions.

The exponential mission reliability function is equal to the exponential survival function evaluated for the failure rate and the mission duration, τ. The exponential mission reliability is a constant over the useful life of the

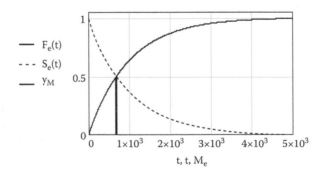

FIGURE A.4
Exponential cumulative failure math model and survival function.

part, since the failure rate and the mission duration are both constant. Taken to the extreme, the reliability of a part characterized by the exponential reliability function is constant for an infinite period of time. The exponential reliability function, for mission duration of eight hours, is expressed in the following equation.

$$R_{\exp}(\tau) = e^{-\lambda\tau} = e^{-(0.00101)(8)} = 0.991953 \tag{A.5}$$

Time-Censored Data

Time-censored data are acquired from empirical investigations (historical and experimental) that have a fixed duration in which not all test articles fail. The total time on test includes the survivors.

1. Time censored without replacement

   Experimental data acquired over a fixed time period in which failed parts are not replaced. For example, an experiment in which the sample number of test articles is able to be tested concurrently, i.e., sufficient test fixtures and test chamber capacity can run all test articles at the same time with no spares.

2. Time censored with replacement

   Experimental data acquired over fixed time period in which failed parts are replaced. For example, an experiment in which the sample number of test articles is not able to be tested concurrently, i.e., test fixtures and test chamber capacity, cannot run all test articles at the same time and failed parts are replaced.

3. Time censored failure free

   Experimental data acquired over fixed time period in which none of the sample test articles fail.

Time Censored without Replacement

Assume that the time-to-failure experiment that was performed was limited to 1,000 hours. All test articles that survived 1,000 hours would be removed from the test without failure. The revised time-censored time-to-failure data are tabulated in Table A.6. Part failures are coded "1" and survivors are coded "C." The number of failures is the sum of the codes where C is not calculated.

The sensor column shows a one for a failure and the letter C for censored data. The total number of failures in the experiment, designated "r," is found by the sum of the censored column, the sum of all of the 1s. The total time on test is the sum of all of the times to failure plus the sum of all of the time on test for the survivors. The point estimate of the MTBF is equal to the total time on tests divided by the total number of failures. We have no

TABLE A.6

Time-Censored Failure Data Without Replacement

TTF	Censored	TTF	Censored
1,000	C	958	1
1,000	C	977	1
1,000	C	1,000	C
1,000	C	994	1
999	1	1,000	C
1,000	C	1,000	C
1,000	C	1,000	C
988	1	1,000	C
1,000	C	989	1
980	1	976	1
966	1	972	1
941	1	992	1
1,000	C	947	1
989	1	975	1
1,000	C	972	1
974	1	996	1
992	1	1,000	C
996	1	978	1
963	1	953	1
972	1	991	1
949	1	987	1
960	1	1,000	C
1,000	C		

knowledge of how long the survivors would remain in operation. Therefore the point estimate of the MTBF is meaningless. The lower confidence limit of the MTBF is calculated using the upper chi-squared distribution, evaluated at one minus the level of significance and $2r + 2$ degrees of freedom. The lower confidence limit is equal to two times the total time on tests divided by the chi-squared distribution. The equation for the lower confidence limit (LCL) of the MTBF is provided below.

$$\theta_{LCL} = \frac{2T}{\chi^2_{1-\alpha,2r+2}} \tag{A.6}$$

Time Censored with Replacement

Assume that the complete data time-to-failure experiment that was performed was limited to 1,000 hours, and replacement test articles were put on test following the failure of the parts that failed. The test data table will include the additional times on test for the replacement test articles, to include possible times to failure, and the subsequent replacement test articles time on test, and a recount for the total number of failures. The total time on test remains the sum of all of the times to failure plus all of the time on test for the survivors. The point estimate of the MTBF is equal to the total time on tests divided by the total number of failures, and the lower confidence limit of the MTBF is the same as above.

Time Censored Failure Free

Assume that all of the parts put on test survived for the 1,000 hours. The total time on test will be equal to the number of test articles times 1,000 hours. The number of failures will be zero. The point estimate of the MTBF is undefined. The lower confidence limit of the MTBF can still be calculated. The chi-squared distribution will have 2 degrees of freedom.

The calculations for the point estimates of the MTBF and the lower confidence limit for the MTBF for the time-censored without replacement and the time-censored failure-free tests are provided in Table A.7.

Weibull Reliability Math Modeling Approach

The Weibull distribution has been recognized as a reliability failure math model for several decades but has been slow to be applied because it is not convenient to fit its parameters from time-to-failure data. The parameters of the Weibull are the scale parameter, η, which is the measure of central tendency, the shape parameter, β, which is the measure of dispersion, and the location parameter, γ, which determines the lower boundary of the distribution. The distribution is often referred to as the two-parameter Weibull,

TABLE A.7

Time-Censored Exponential Failure Math Model with LC for Means

	Math Model	
	r Failures	Failure Free
θ	1,583.07	#DIV/0!
λ	0.000632	#DIV/0!
α	0.05	0.05
r	28	0
ν	58	2
$\chi^2_{\alpha,\nu}$	76.78	5.99
t	44,326	45,000
θ_{LCL}	1,154.66	15,021.37
λ_{LCL}	0.000866	0.000067

when $\gamma = 0$, and the three-parameter Weibull, when $\gamma > 0$. The Weibull is actually a family of distributions that range from parabolic, when $0 < \gamma < 1$; to exponential, when $\gamma = 1$; to positively skewed, when $1 < \gamma < 3.6$; to symmetrical, when $\gamma = 3.6$; and to negatively skewed, when $\gamma > 3.6$.

Complete Data

The procedure for fitting time-to-failure and time-to-repair data to the Weibull distribution is presented in Excel in the following steps:

1. Fit data to the Weibull distribution for location parameter equal to zero using the median regression method (often referred to as the two-parameter Weibull distribution).
2. Find the best-fit value for the location parameter using correlation analysis.
3. Fit Weibull distribution for best-fit value for the location parameter.

The procedure is

1. Rank order the time-to-failure data from lowest to highest (Table A.8).
2. Assign an index number to the rank order data from 1 to n (Table A.9).
3. Calculate Bartlett's median ranks estimator for the cumulative distribution, F, for each index number (Table A.10) where

$$F = \frac{i - 0.3}{n + 0.4} \tag{A.7}$$

4. Calculate the independent variable for the median ranks regression, $X\gamma$, for each time to failure, TTF_i (Table A.11), where

$$X_\gamma = \ln(TTF_i - \gamma) \tag{A.8}$$

Let $\gamma = 0$ for the initial calculation of $X\gamma$.

TABLE A.8	
Rank Order TTF	
TTF$_i$	
Raw	Ranked
526	520
520	522
522	525
525	526

TABLE A.9		
Assign Index Number		
TTF$_i$		
Raw	Ranked	i
526	520	1
520	522	2
522	525	3
525	526	4

TABLE A.10

Calculate Bartlett's Median Ranks

TTF$_i$			
Raw	**Ranked**	***i***	***F***
526	520	1	0.159
520	522	2	0.386
522	525	3	0.614
525	526	4	0.841

Note: $i = 1 \rightarrow F = (1 - 0.3)(n + 0.4) = 1.3/4.4$

TABLE A.11

Calculate Xγ

TTF$_i$				
Raw	**Ranked**	***i***	***F***	**Xγ**
526	520	1	0.159	6.254
520	522	2	0.386	6.258
522	525	3	0.614	6.263
525	526	4	0.841	6.265

Note: $i = 1 \rightarrow X\gamma = \ln(TTF_i - \gamma) = \ln(520 - 0)$

TABLE A.12

Calculate Y

TTF$_i$					
Raw	**Ranked**	***i***	***F***	**Xγ**	**Y**
526	520	1	0.159	6.254	−1.753
520	522	2	0.386	6.258	−0.717
522	525	3	0.614	6.263	−0.050
525	526	4	0.841	6.265	0.609

Note: $i = 1 \rightarrow Y = \ln(\ln(1/(1 - F))) = \ln(\ln(1/(1 - 0.159)))$

5. Calculate the dependent variable for the median ranks regression, Y, for each time to failure, TTF$_i$ (Table A.12), where

$$Y = \ln(\ln(1/(1 - F))) \tag{A.9}$$

6. Calculate the coefficient of determination, r^2, for the median ranks regression (Table A.13). The Excel data analysis correlation calculator is applied to the Xγ and Y columns. Coefficient of correlation squared is equal to the coefficient of determination.

The coefficient of determination is the measure of how well the data are fit to the median ranks regression.

TABLE A.13

Calculate r^2 for $\gamma = 0$

X_γ	Y		X_γ	Y
			$\gamma = 0$	
6.254	−1.75	X_γ	1	
6.258	−0.72	Y	0.98254	1
6.263	−0.05	$r^2 =$	0.96538	
6.265	0.609			

TABLE A.14

Initial Estimate for Location Parameter

X_γ	Y		X_γ	Y
			$\gamma = 250$	
5.598	−1.75	X_γ	1	
5.606	−0.72	Y	0.98261	1
5.617	−0.05	$r^2 =$	0.96552	
5.62	0.609			

7. Set the initial value of γ to a value between 0 and the minimum TTF (250 hr) (Table A.14)

8. Calculate the coefficient of determination for the median ranks regression using $X_\gamma - \gamma$.

   a. If the coefficient of correlation is less than that for $\gamma = 0$, then recalculate the median ranks regression for $\gamma = 0$, and run the Excel data analysis regression analysis.

   b. If the coefficient of correlation is greater than that for $\gamma = 0$, then iteratively increase the values for γ until the coefficient of determination is maximized (Table A.15), set the value for X_γ at that value of γ, and run the Excel data analysis regression analysis (Table A.16).

9. The regression table will typically include a table of regression statistics, an analysis of variance (ANOVA) table, and a table of coefficients. The coefficient of X_γ is the slope of the regression model and is equal to the shape parameter, β. The scale parameter, h, is equal to the anti-log of the ratio of the y-intercept of the regression model and the shape parameter (Table A.17).

$$\eta = \ln^{-1}\left(\frac{y_0}{\beta}\right) = e^{-\left(\frac{y_0}{\beta}\right)} \tag{A.10}$$

TABLE A.15

Iterative Approach to Optimize
Location Parameter

γ	r^2	$\gamma = 510$	$X\gamma$	Y
0	0.96538	$X\gamma$	1	
250	0.96552	Y	0.98487	1
510	0.96996		$r^2 = 0.96996$	
512	0.97045	$\gamma = 512$	$X\gamma$	Y
513	0.96964	$X\gamma$	1	
		Y	0.98511	1
			$r^2 = 0.97045$	
		$\gamma = 513$	$X\gamma$	Y
		$X\gamma$	1	
		Y	0.98470	1
			$r^2 = 0.96964$	

TABLE A.16

Median Ranks Table for $X\gamma$ and Y
at Optimum Location Parameter

$\gamma = 512$	
$X\gamma$	Y
2.079	−1.753
2.303	−0.717
2.565	−0.050
2.639	0.609

The procedure is illustrated for the Weibull reliability failure model apply-
ing the median ranks regression approach using the time-to-failure complete data from the exponential distribution restated in Table A.18.

Median Ranks Regression Table

- The time-to-failure data, TTF, is rank ordered in the first column.
- An index number ranging from 1 to n is assigned to each data point.
- Bartlett's median ranks estimator is calculating for the cumulative
  probability distribution, F.
- The independent variable, $X\gamma$, is calculated for $\gamma = 0$.
- the dependent variable, Y, is calculated for i and n.

TABLE A.17

Median Ranks Regression Method to Fit Parameters of Weibull Distribution

Summary Output

Regression Statistics

Multiple R	0.985
R^2	0.970
Adj R^2	0.956
Standard error	0.212
Observations	4

ANOVA	df	SS	MS	F	P		
Regression	1	2.956	2.956	65.677	0.015		
Residual	2	0.090	0.045				
Total	3	3.046					
Coefficients		SE	t	P	LCI 95%	UCI 95%
Intercept: y_0	−9.772	1.152	−8.484	0.014	−14.728	−4.817
$X\gamma - \beta a$	3.878	0.479	8.104	0.015	1.819	5.937
η	12.425	exp $(-y_0/\beta)$				

TABLE A.18

Median Ranks Regression Input Data

TTF	i	F	$X\gamma$	Y
941	1	0.015	3.045	−4.164
947	2	0.037	3.296	−3.266
949	3	0.059	3.367	−2.792
953	4	0.081	3.497	−2.465
958	5	0.104	3.638	−2.214
960	6	0.126	3.689	−2.009
963	7	0.148	3.761	−1.835
966	8	0.170	3.829	−1.683
972	9	0.192	3.951	−1.548
972	10	0.214	3.951	−1.426
972	11	0.236	3.951	−1.314
974	12	0.258	3.989	−1.211
975	13	0.280	4.007	−1.114
976	14	0.302	4.025	−1.024
977	15	0.324	4.043	−0.938
978	16	0.346	4.060	−0.857
980	17	0.368	4.094	−0.780
987	18	0.390	4.205	−0.705
988	19	0.412	4.220	−0.633
989	20	0.434	4.234	−0.564

TABLE A.18 (continued)

Median Ranks Regression Input Data

TTF	i	F	$X\gamma$	Y
989	21	0.456	4.234	−0.496
991	22	0.478	4.263	−0.431
992	23	0.500	4.277	−0.367
992	24	0.522	4.277	−0.304
994	25	0.544	4.304	−0.242
996	26	0.566	4.331	−0.180
996	27	0.588	4.331	−0.120
999	28	0.610	4.369	−0.060
1,001	29	0.632	4.394	0.000
1,002	30	0.654	4.407	0.060
1,002	31	0.676	4.407	0.120
1,004	32	0.698	4.431	0.181
1,006	33	0.720	4.454	0.242
1,006	34	0.742	4.454	0.304
1,006	35	0.764	4.454	0.368
1,007	36	0.786	4.466	0.434
1,007	37	0.808	4.466	0.502
1,016	38	0.830	4.564	0.573
1,016	39	0.852	4.564	0.649
1,023	40	0.874	1.635	0.730
1,023	41	0.896	4.635	0.819
1,024	42	0.919	4.644	0.919
1,026	43	0.941	4.663	1.038
1,032	44	0.963	4.718	1.189
1,037	45	0.985	4.762	1.428

Finding the Location Parameter, γ

- The initial value for the coefficient of determination, r^2, is calculated from the coefficient correlation routine in Excel, 0.96082.
- An estimate for γ between zero and the minimum TTF, 500 hours, is used to recalculate the values of the independent variable X gamma.
- The coefficient of determination for $\gamma = 500$ hours increases to 0.96520, greater than the coefficient of determination for $\gamma = 0$. This tells us that the Weibull distribution with a location parameter of 500 hours is a better fit to the data than a location parameter of zero hours. The search for the best fit for the Weibull distribution is an iterative process to find the value of the location parameter that has the highest coefficient of determination.
- The best-fit estimate for location parameter is found to be equal to 920 hours, as shown in Table A.19.

TABLE A.19

Coefficient of Determination for Various Values of the Location Parameter

γ	r^2		$X\gamma$	Y
0	0.96082	$X\gamma$	1	
500	0.96520	Y	0.99567	1
750	0.97336	$r^2 =$	0.99136	
850	0.98278			
900	0.99090			
920	0.99136			
925	0.98913			
930	0.98354			

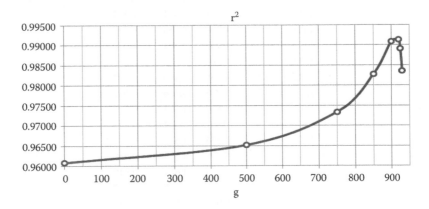

FIGURE A.5
Coefficient of determination plot.

The search for the best estimate for the location parameter is illustrated in Figure A.5.

- The median ranks regression table is reevaluated for the independent variable, $X\gamma$, for $\gamma = 920$ hours.
- The Excel regression routine is run for the independent and dependent variables of the median ranks regression, and the summary output is provided in Table A.20.
- The regression statistics confirm the coefficient of determination found in the search for the best-fit estimate of the location parameter.
- The ANOVA table rejects the null hypothesis that there is no statistically significant difference between values of the dependent independent variable and confirms that a statistically significant regression model exists.

TABLE A.20

Median Ranks Regression and Parameters of Weibull Failure Math Model

Regression Statistics					
Multiple R	0.996				
R^2	0.991				
Adjusted R^2	0.991				
Standard Error	0.114				
Observations	45				

ANOVA	df	SS	MS	F	P
Regression	1	64.58	64.58	4,935	0.000
Residual	43	0.56	0.01		
Total	44	65.14			

Coefficients		SE	t	P	LCI 95%	UCI 95%
y_0	−13.42	0.18	−72.99	0.000	−13.79	−13.05
β	3.07	0.04	70.25	0.000	2.99	3.16
η	78.87					
γ	920					

- The coefficients table provides two inferences:
  - Neither the y-intercept nor the slope of the regression model are statistically the same as zero.
  - The shape parameter of the Weibull distribution, the slope of the regression model, $\beta = 3.07$, and the scale parameter of the Weibull distribution can be calculated as the analog of the ratio of the y-intercept to the slope, $\eta = 78.87$ hours.

A graphical solution to the median ranks regression for the parameters of the Weibull distribution is presented in Figure A.6. The scattergram represents the locations of the independent and dependent variables. The trend line routine in Excel plots a linear least squares regression line with the equation and the coefficient of determination. Determination of the shape and scale parameters for the Weibull is performed in the same manner as from the regression analysis. The only information that is lacking is the confidence interval for each parameter to determine whether it is statistically the same as zero.

The Weibull failure math model for the complete data time to failure (Figure A.7) is expressed as

$$f(t) = \begin{vmatrix} 0 & \text{if} \quad t < 920 \\ \left(\dfrac{3.07}{78.87^{3.07}}\right)(t-920)^{2.07} \, e^{-\left(\frac{t-920}{78.87}\right)^{3.07}} & \text{if} \quad t \geq 920 \end{vmatrix} \qquad (A.11)$$

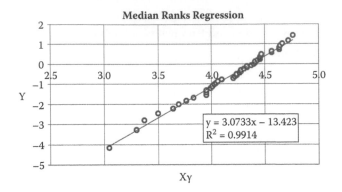

FIGURE A.6
Graphical approach median ranks regression and parameters of Weibull failure math model.

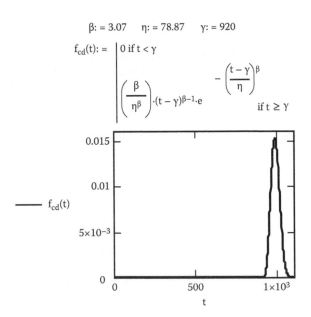

FIGURE A.7
Weibull failure math model.

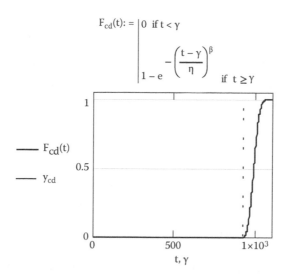

$$F_{cd}(t): = \begin{vmatrix} 0 & \text{if } t < \gamma \\ \\ 1 - e^{-\left(\frac{t-\gamma}{\eta}\right)^{\beta}} & \text{if } t \geq \gamma \end{vmatrix}$$

FIGURE A.8
Weibull cumulative failure math model.

The Weibull cumulative failure math model (Figure A.8) is expressed as

$$F(t) = \begin{vmatrix} 0 & \text{if} & t < 920 \\ \\ 1 - e^{-\left(\frac{t-920}{78.87}\right)^{3.07}} & \text{if} & t \geq 920 \end{vmatrix}$$

(A.12)

The Weibull median time to failure is expressed as

$$M_W = \gamma + \eta \left(\ln(2) \right)^{\frac{1}{\beta}}$$

$$M_W = 920 + 78.84 \left(\ln(2) \right)^{\frac{1}{3.07}}$$

(A.13)

The cumulative probability distribution for the Weibull failure math model showing the location of the median is presented in Figure A.9.

The expression for the Weibull survival function is provided in the following equation.

$$S_W(t) = \begin{vmatrix} 1 & \text{if} & t < 920 \\ \\ e^{-\left(\frac{t-920}{78.87}\right)^{3.07}} & \text{if} & t \geq 920 \end{vmatrix}$$

(A.14)

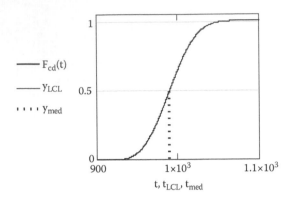

FIGURE A.9
Weibull cumulative failure math model with median.

The expression for the Weibull mission reliability function for a mission duration, τ, of 16 hours is provided in the following equation

$$R_W(\tau \mid t) = \begin{vmatrix} 1 & \text{if} & t < 920 \\ \dfrac{e^{-\left(\frac{t-920+16}{78.87}\right)^{3.07}}}{e^{-\left(\frac{t-920}{78.87}\right)^{3.07}}} & \text{if} & t \ge 920 \end{vmatrix} \tag{A.15}$$

The Weibull survival, heavy solid black line, and mission reliability, light solid black line, functions are plotted in Figure A.10. The value of a location parameter is plotted in the dashed line. The survival and mission reliability

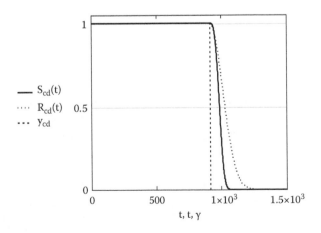

FIGURE A.10
Weibull survival and mission reliability functions.

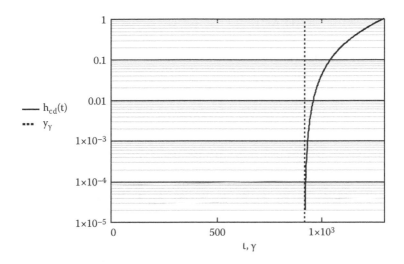

FIGURE A.11
Weibull hazard function.

functions are equal to one until they reached the location parameter, at which point they decline and approach zero.

The Weibull hazard function, instantaneous failure rate, is expressed in the following equation

$$
h_W(t) = \begin{vmatrix} 0 & \text{if} & t < 920 \\ \left(\dfrac{3.07}{78.87^{3.07}}\right)(t-920)^{3.07-1} & \text{if} & t \geq 920 \end{vmatrix}
\tag{A.16}
$$

The Weibull hazard function is plotted in Figure A.11. Notice that the instantaneous failure rate is 0 for all operating hours less than the location parameter.

Time-Censored Data

Consider the time-censored data from the example for the exponential distribution, provided in Table A.21. The time-to-failure data for failed parts is coded 1, the surviving parts are coded C.

- The procedure for the median ranks regression analysis assigns an index number to all of the parts. The calculation of the Bartlett's median ranks estimator for the cumulative probability distribution uses the value of the index for each part and the value of n representing all of the parts.

TABLE A.21

Median Ranks Regression Input
for Failure Censored Data

TTF	Cen	i	F	$X\gamma$	Y
941	1	1	0.015	2.773	−4.164
947	1	2	0.037	3.091	−3.266
949	1	3	0.059	3.178	−2.792
953	1	4	0.081	3.332	−2.465
958	1	5	0.104	3.497	−2.214
960	1	6	0.126	3.555	−2.009
963	1	7	0.148	3.638	−1.835
966	1	8	0.170	3.714	−1.683
972	1	9	0.192	3.850	−1.548
972	1	10	0.214	3.850	−1.426
972	1	11	0.236	3.850	−1.314
974	1	12	0.258	3.892	−1.211
975	1	13	0.280	3.912	−1.114
976	1	14	0.302	3.932	−1.024
977	1	15	0.324	3.951	−0.938
978	1	16	0.346	3.970	−0.857
980	1	17	0.368	4.007	−0.780
987	1	18	0.390	4.127	−0.705
988	1	19	0.412	4.143	−0.633
989	1	20	0.434	4.159	−0.564
989	1	21	0.456	4.159	−0.496
991	1	22	0.478	4.190	−0.431
992	1	23	0.500	4.205	−0.367
992	1	24	0.522	4.205	−0.304
994	1	25	0.544	4.234	−0.242
996	1	26	0.566	4.263	−0.180
996	1	27	0.588	4.263	−0.120
999	1	28	0.610	4.304	−0.060
1,000	C	29	0.632	4.317	0.000
1,000	C	30	0.654	4.317	0.060
1,000	C	31	0.676	4.317	0.120
1,000	C	32	0.698	4.317	0.181
1,000	C	33	0.720	4.317	0.242
1,000	C	34	0.742	4.317	0.304
1,000	C	35	0.764	4.317	0.368
1,000	C	36	0.786	4.317	0.434
1,000	C	37	0.808	4.317	0.502
1,000	C	38	0.830	4.317	0.573

TABLE A.21 (continued)

Median Ranks Regression Input
for Failure Censored Data

TTF	Cen	i	F	$X\gamma$	Y
1,000	C	39	0.852	4.317	0.649
1,000	C	40	0.874	4.317	0.730
1,000	C	41	0.896	4.317	0.819
1,000	C	42	0.919	4.317	0.919
1,000	C	43	0.941	4.317	1.038
1,000	C	44	0.963	4.317	1.189
1,000	C	45	0.985	4.317	1.428

- The value for the independent variable, $X\gamma$, is calculated in the same fashion as complete data.
- The value for the dependent variable, Y, is also calculated in the same fashion; notice that its values are only relevant for time-to-failure data for failed parts. (In a time-censored data experiment with replacements, many of the censored surviving time on test would be included among the failed parts' time-to-failure data.)
- Search for the best-fit estimate for the location parameter is performed only for the 28 time-to-failure data, the censored data are excluded. The best-fit estimate for the location parameter is found to be 925 hours as shown in Table A.22.
- The median ranks regression table is reevaluated for the independent variable, $X\gamma$, for $\gamma = 925$ hours.

TABLE A.22

Failure Censored Coefficient
of Determination for Various
Values of the Location Parameter

γ	r^2		$X\gamma$	Y
0	0.95512	$X\gamma$	1	
500	0.95802	Y	0.99515	1
750	0.96405	r^2	0.99033	
850	0.97247			
900	0.98310			
920	0.98944			
925	0.99033			
930	0.98930			

TABLE A.23

Median Ranks Regression and Parameters for Failure Censored Weibull Failure Math Model

Regression Statistics					
Multiple R	0.995				
R^2	0.990				
Adjusted R^2	0.990				
Standard Error	0.102				
Observations	28				

ANOVA	df	SS	MS	F	P
Regression	1	27.76	27.76	2,662.29	0.000
Residual	26	0.27	0.01		
Total	27	28.03			

Coefficients		SE	t	P	LCI 95%	UCI 95%
y_0	−11.18	0.19	−57.75	0.000	−11.57	−10.78
β	2.57	0.05	51.60	0.000	2.47	2.67
η	77.36					
γ	925					

- The Excel regression routine is run for the independent and dependent variables of the median ranks regression, and the summary output is provided in Table A.23.
- The regression statistics confirm the coefficient of determination found in the search for the best-fit estimate of the location parameter.
- The ANOVA table rejects the null hypothesis that there is no statistically significant difference between values of the dependent independent variable and confirms that a statistically significant regression model exists.
- The coefficients table provides two inferences:
  - Neither the y-intercept nor the slope of the regression model are statistically the same as zero.
  - The shape parameter of the Weibull distribution; the slope of the regression model, β = 2.57; and the scale parameter of the Weibull distribution can be calculated as the analog of the ratio of the y-intercept to the slope, η = 77.36 hours.

Comparison of the two data sets observes relatively small changes in the magnitude of the parameters of the Weibull distribution.

The graphical solution to the median ranks regression for the time-censored data is presented in Figure A.12.

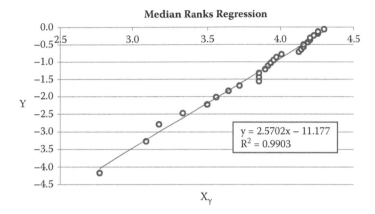

FIGURE A.12
Graphical approach median ranks regression and parameters for failure censored Weibull failure math model.

The Weibull distribution reliability functions are fit with the time-censored estimates for the shape, scale, and location parameters in Figures A.13–A.16. The reliability functions follow:

Failure math model

$$f(t) = \begin{vmatrix} 0 & \text{if} & t < 925 \\ \left(\dfrac{2.57}{77.36^{2.57}}\right)(t-925)^{1.57}\, e^{-\left(\frac{t-925}{77.36}\right)^{2.57}} & \text{if} & t \geq 925 \end{vmatrix} \qquad (A.17)$$

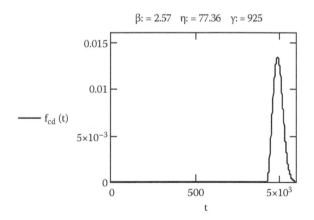

FIGURE A.13
Failure-censored Weibull failure math model.

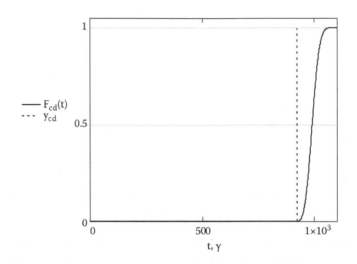

FIGURE A.14
Failure-censored cumulative Weibull failure math model.

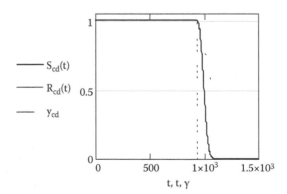

FIGURE A.15
Failure-censored Weibull mission reliability and survival functions.

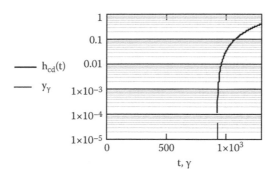

FIGURE A.16
Failure-censored Weibull hazard function.

Cumulative failure math model

$$F(t) = \begin{vmatrix} 0 & \text{if} & t < 925 \\ 1 - e^{-\left(\frac{t-925}{77.36}\right)^{2.57}} & \text{if} & t \geq 925 \end{vmatrix} \tag{A.18}$$

Survival function

$$S(t) = \begin{vmatrix} 1 & \text{if} & t < 925 \\ e^{-\left(\frac{t-925}{77.36}\right)^{2.57}} & \text{if} & t \geq 925 \end{vmatrix} \tag{A.19}$$

Mission reliability function

$$R(t) = \begin{vmatrix} 1 & \text{if} & t < 925 \\ \dfrac{e^{-\left(\frac{t-925+16}{77.36}\right)^{2.57}}}{e^{-\left(\frac{t-925}{77.36}\right)^{2.57}}} & \text{if} & t \geq 925 \end{vmatrix} \tag{A.20}$$

Hazard function

$$h(t) = \begin{vmatrix} 0 & \text{if} & t < 925 \\ \left(\dfrac{2.57}{77.36^{2.57}}\right)(t-925)^{1.57} & \text{if} & t \geq 925 \end{vmatrix} \tag{A.21}$$

Comparative Evaluation between Exponential and Weibull Failure Math Modeling Approaches

Is important understand how the exponential and Weibull failure math models describe the time-to-failure data from the experiment. The objective of course is to find a continuous distribution that best fits the data. A histogram of the sample time-to-failure data is provided in Figure A.17 with the graphs for the respective failure math models.

It is readily apparent from the graph that the Weibull failure math model fits the histogram of the time-to-failure data and the exponential distribution does not. Another example of the distinction between the two distributions is the cumulative failure math model, as shown in Figure A.18. The sample data suggest that the part will function failure free until the operating time reaches a location parameter, at which time the parts fail as shown in the Weibull failure math model. The Weibull cumulative failure math model begins with the part from zero following the occurrence of the location parameter and appears to have approached 100% at the maximum range of the sample data. The exponential cumulative failure math model suggests that at least 60% of the parts will have failed before the location parameter is reached. Although the sample data show no failures in the range from 0 to 500 hours, the exponential cumulative failure math model suggests that at least 40% of the parts will have failed by 500 hours.

The exponential and Weibull mission reliability functions are provided in Figure A.19. The exponential mission reliability function is a constant over

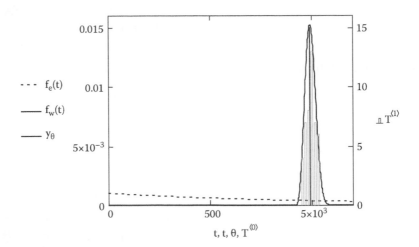

FIGURE A.17
Exponential and Weibull failure math models.

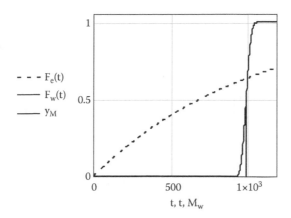

FIGURE A.18
Exponential and Weibull cumulative failure math models.

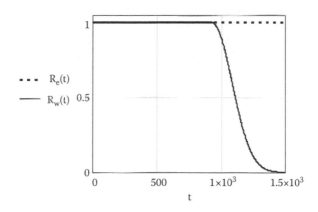

FIGURE A.19
Exponential and Weibull mission reliability functions.

the range $0 < t <$ infinity, while the Weibull mission reliability function suggests that the part will experience a failure-free duration with a reliability equal to one until the part reaches the age of the location parameter. After the part reaches the age of the location parameter the reliability of the part will decline as the part wears out.

The exponential and Weibull hazard functions are provided in Figure A.20. The exponential hazard function is a constant over the useful life of the part, again ranging from $0 < t <$ infinity. The Weibull hazard function suggests that the part will experience a failure-free period, at which point

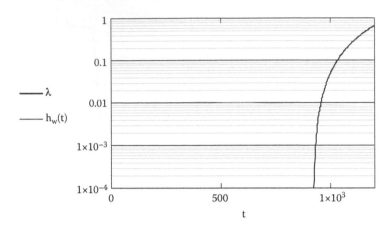

FIGURE A.20
Exponential and Weibull hazard functions.

the risk of failure, the instantaneous failure rate, will increase as the part ages beyond the location parameter.

I leave it to the reader to determine which of the two approaches, exponential or Weibull failure analysis, provide the best analytical estimation of the true behavior of part failure.

Appendix B: Maintainability Math Models and Maintainability Functions

1. Time-to-repair math model, log-normal distribution approach
2. Time-to-repair math model, Weibull distribution approach

Time-to-Repair Math Model, Log-Normal Distribution Approach

The best practice for fitting the repair math models from empirical data applies the log-normal distribution to account for the typically positively skewed time-to-repair data. Analysts recognized that the data could not be evaluated using the standard normal probability distribution.

The example for fitting the repair math models will use Table B.1 for time-to-repair data. Notice that the units of time are expressed in minutes. Repair experiments measure the time to perform a list of work instructions for tasks that often last less than eight hours. Attempts to record the raw data in hours provides an opportunity for error that can be mitigated by using minutes as your time metric. The two critical metrics for repair are the mean time to repair (MTTR) and its upper confidence interval (UCI), which are expressed in units of hours. Dividing the final results of an analysis performed in minutes by 60 minutes per hour enables us to report MTTR and its upper confidence limit in hours without the possibility of error.

The descriptive statistics for the time-to-repair data are provided in Table B.2. The MTTR is 172 minutes. But we observe that the kurtosis (2.304) and the skewness (1.027) describe a sample that has a much higher peak than the normal distribution and is significantly skewed to the right.

A frequency distribution for the time-to-repair data is provided in Table B.3. Sample frequency distributions for continuous random variables provide a rough estimate for the shape of the distribution. The selection of the number of class intervals and their width can make a visual assessment of the shape of the distribution difficult. If the sample frequency distribution in Table B.3 used five one-hour class intervals, the frequency distribution would have three values equal to 30, 9, 1. If the sample frequency distribution in Table B.3 used 27 ten-minute class intervals, the frequency distribution would have appeared to be flat verging on uniform.

TABLE B.1

Time-to-Repair Data in Minutes

186	131	192	166	159
131	195	155	164	169
222	174	154	168	179
179	153	177	153	159
158	247	158	149	189
163	171	176	184	179
206	206	184	146	179
175	150	168	169	159

TABLE B.2

Time-to-Repair Descriptive Statistics

Mean	172.05
Standard Error	3.57
Median	169
Mode	179
Standard Deviation	22.59
Sample Variance	510.41
Kurtosis	2.304
Skewness	1.027
Range	116
Minimum	131
Maximum	247
Sum	6,882
Count	40

TABLE B.3

Time-to-Repair Sample
Frequency Distribution

TTR	TTR	$f_s(t)$
0	0	0
30	30	0
60	60	0
90	90	0
120	120	0
150	150	5
180	180	25
210	210	8
240	240	1
270	270	1

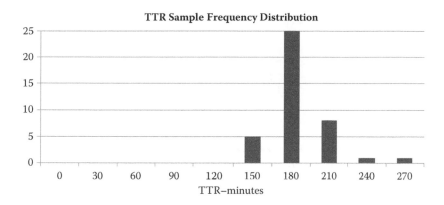

FIGURE B.1
Time-to-repair sample frequency distribution histogram.

The sample frequency distribution chosen by the selection of the number and width of the class intervals provides a histogram as shown in Figure B.1. The high peak at 180 minutes and the declining data to the right confirm the expectation of the shape of the distribution as expressed by the kurtosis in the skewness.

To provide an example of how a sample frequency distribution might mislead the analyst in attempting to understand the shape of the continuous distribution, 15-minutes class widths were applied to the TTR data as shown in Table B.4.

Plotting the sample frequency distribution histogram reveals a shape that almost appears to be symmetrical about 180 minutes and may not be perceived as skewed (Figure B.2). One might consider this to be an exercise in deceit used by unscrupulous people who use valid data to draw an invalid premise.

TABLE B.4

Sample Frequency Distribution With 15 Minute Class Widths

TTR	TTR	$f_s(t)$
120	120	0
135	135	2
150	150	3
165	165	11
180	180	14
195	195	6
210	210	2
225	225	1
240	240	0
255	255	1

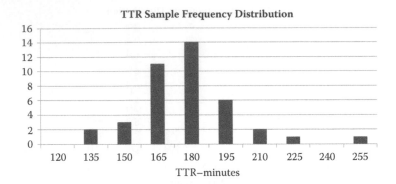

FIGURE B.2
Time-to-repair sample frequency distribution histogram with 15-minute class width.

The procedure for fitting the log-normal repair math model begins with taking the natural logarithm of the time-to-repair data as shown in Table B.5.

The descriptive statistics are now expressed in the log of time to repair. Inferences for the upper confidence limit of the time to repair can apply the same procedure used for normal data, the upper confidence limit equals the mean plus the product of the Student's t distribution evaluated at $(1 - \alpha)$ and $v = n - 1$ degrees of freedom and the standard deviation (Table B.6).

The upper confidence limit for time to repair is readily calculated for our example sample data in two steps.

1. Calculate the upper confidence limit for the log of the time-to-repair data.

$$\text{UCL}_{\ln \text{TTR}} = \mu_{\ln \text{TTR}} + t_{(1-\alpha, v)} \sigma_{\ln \text{TTR}} \tag{B.1}$$

2. Convert the upper confidence limit for the log of the time-to-repair data to the upper confidence limit in hours.

The antilog of the upper confidence limit for the log of the mean-time-to-repair data provides the upper confidence limit in hours. The antilog is found as

$$\text{UCL}_{\text{minutes}} = e^{\text{UCL}_{\ln \text{TTR}}} \tag{B.2}$$

The upper confidence limit for the MTTR is calculated in the same procedure except the standard error (SE) of the mean (SEM) is used instead of the standard deviation. Because of the central limit theorem, confidence limits for means can be calculated without regard to the distribution from which the

TABLE B.5

Time-to-Repair Data Natural
Log Transformation

TTR	ln (TTR)
3.10	1.131
2.18	0.781
3.70	1.308
2.98	1.093
2.63	0.968
2.72	0.999
3.43	1.234
2.92	1.070
2.18	0.781
3.25	1.179
2.90	1.065
2.55	0.936
4.12	1.415
2.85	1.047
3.43	1.234
2.50	0.916
3.20	1.163
2.58	0.949
2.57	0.943
2.95	1.082
2.63	0.968
2.93	1.076
3.07	1.121
2.80	1.030
2.77	1.018
2.73	1.006
2.80	1.030
2.55	0.936
2.48	0.910
3.07	1.121
2.43	0.889
2.82	1.036
2.65	0.975
2.82	1.036
2.98	1.093
2.65	0.975
3.15	1.147
2.98	1.093
2.98	1.093
2.65	0.975

TABLE B.6

Descriptive Statistics for Transformed
TTR Data

ln (TTR)	
Mean	1.045
Standard Error	0.020
Median	1.036
Mode	1.093
Standard Deviation	0.127
Sample Variance	0.016
Kurtosis	1.221
Skewness	0.488
Range	0.634
Minimum	0.781
Maximum	1.415
Sum	41.819
Count	40

data are derived. It was not necessary to calculate the natural logarithm of
the time-to-repair data to calculate the upper confidence limit for the mean.

1. Calculate the upper confidence limit for the log of the time-to-
   repair data.

$$\text{UCL}_{1\mu_{\ln TTR}} = \mu_{\ln TTR} + t_{(1-\alpha,v)}\text{SEM}_{\ln TTR} \tag{B.3}$$

2. Convert the upper confidence limit (Table B.7) for the log of the time-
   to-repair data to the upper confidence limit in hours.

$$\text{UCL}_{\mu_{minutes}} = e^{\text{UCL}_{\mu_{\ln TTR}}} \tag{B.4}$$

TABLE B.7

Upper Confidence Limits
of Mean and Sample TTR

μ	1.045	
SEM	0.020	
s	0.127	
α	0.05	
v	39	
$t_{\alpha,v}$	1.685	
UCL_μ (hr)	1.079	2.94
UCL_s (hr)	1.259	3.52

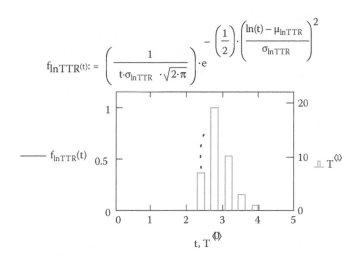

FIGURE B.3
Log-normal distribution for TTR sample data.

The log-normal repair math model is expressed in the following equation. Notice that it differs from the expression for the normal distribution in the first factor. The denominator of the first factor includes the independent variable, t.

$$f_{\ln TTR}(t) = \left(\frac{1}{t * \sigma_{\ln TTR} * \sqrt{2\pi}} \right) * e^{-\left(\frac{1}{2}\right)\left(\frac{\ln(t) - \mu_{\ln TTR}}{\sigma_{\ln TTR}}\right)^2} \tag{B.5}$$

The continuous repair math model is provided in Figure B.3 and shows the relationship with the histogram of the time-to-failure data.

Time-to-Repair Math Model, Weibull Distribution Approach

The detailed procedure for fitting data to the Weibull probability mass density function is provided in Appendix A using time-to-failure data. The time-to-repair data are presented without discussion in Tables B.8–B.11 and Figures B.4–B.6.

TABLE B.8

Median Ranks Regression Input Table for TTR
Data and Optimum Value of Location Parameter

TTR	i	F	$X\gamma$	Y
2.18	1	0.017	−1.261	−4.047
2.18	2	0.042	−1.261	−3.147
2.43	3	0.067	−0.629	−2.671
2.48	4	0.092	−0.539	−2.343
2.50	5	0.116	−0.511	−2.090
2.55	6	0.141	−0.431	−1.883
2.55	7	0.166	−0.431	−1.707
2.57	8	0.191	−0.405	−1.554
2.58	9	0.215	−0.381	−1.417
2.63	10	0.240	−0.310	−1.293
2.63	11	0.265	−0.310	−1.179
2.65	12	0.290	−0.288	−1.073
2.65	13	0.314	−0.288	−0.974
2.65	14	0.339	−0.288	−0.881
2.72	15	0.364	−0.203	−0.793
2.73	16	0.389	−0.182	−0.709
2.77	17	0.413	−0.143	−0.629
2.80	18	0.438	−0.105	−0.551
2.80	19	0.463	−0.105	−0.476
2.82	20	0.488	−0.087	−0.402
2.82	21	0.512	−0.087	−0.331
2.85	22	0.537	−0.051	−0.261
2.90	23	0.562	0.000	−0.192
2.92	24	0.587	0.017	−0.124
2.93	25	0.611	0.033	−0.056
2.95	26	0.636	0.049	0.011
2.98	27	0.661	0.080	0.078
2.98	28	0.686	0.080	0.146
2.98	29	0.710	0.080	0.214
2.98	30	0.735	0.080	0.284
3.07	31	0.760	0.154	0.355
3.07	32	0.785	0.154	0.429
3.10	33	0.809	0.182	0.505
3.15	34	0.834	0.223	0.586
3.20	35	0.859	0.262	0.672
3.25	36	0.884	0.300	0.766
3.43	37	0.908	0.427	0.872
3.43	38	0.933	0.427	0.995
3.70	39	0.958	0.588	1.153
4.12	40	0.983	0.796	1.400

TABLE B.9

Coefficient of Determination for Values
of Location Parameter

γ	r^2		$X\gamma$	Y
0	0.89026	$X\gamma$	1	
1.00	0.91686	Y	0.97627	1
1.10	0.92065	r^2	0.95310	
1.20	0.92471			
1.30	0.92905			
1.40	0.93366			
1.50	0.93849			
1.60	0.94342			
1.70	0.94815			
1.80	0.95197			
1.90	0.95310			
1.95	0.95133			
2.00	0.94633			

Note: The r^2 value when γ is 1.90 is equal to
the r^2 value for $X\gamma$.

TABLE B.10

Median Ranks Regression for Parameters of Weibull Distribution

Regression Statistics

Multiple R	0.976
R^2	0.953
Adjusted R^2	0.952
Standard Error	0.266
Observations	40

ANOVA	df	SS	MS	F	P
Regression	1	54.55	54.55	772.22	0.000
Residual	38	2.68	0.07		
Total	39	57.24			

Coefficients		SE	t	P	LCI 95%	UCI 95%
y_0	−0.244	0.04	−5.60	0.000	−0.332	−0.156
β	2.881	0.10	27.79	0.000	2.671	3.091
η	1.088					
γ	1.900					

TABLE B.11

Excel Goal Seek for Weibull
TTR Upper Confidence Limit

C	0.95
UCL	3.49
F(UCL)	0.95

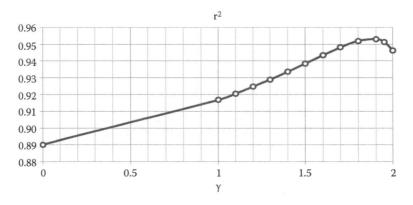

FIGURE B.4
Coefficient of determination plot.

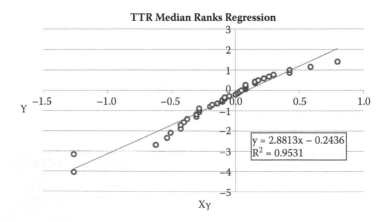

FIGURE B.5
Graphical approach median ranks regression for parameters of Weibull distribution.

$$f_w(t) := \begin{vmatrix} 0 \text{ if } t < \gamma \\ \left(\dfrac{\beta}{\eta^\beta}\right) \cdot (t - \gamma)^{\beta-1} \cdot e^{-\left(\frac{t-\gamma}{\eta}\right)^\beta} \text{ if } t \geq \gamma \end{vmatrix}$$

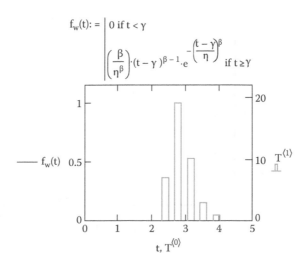

FIGURE B.6
Weibull repair math model.

Comparison between the Log-Normal and Weibull Approach

Figure B.7 represents the distinctions between how well the data fits the log-normal (gray dashed lines) and Weibull (bold black lines) distributions. Recall that the Weibull distribution will give the exact results provided by the log-normal distribution when the shape parameter is equal to approximately 2.26. When the shape parameter is any value other than 2.26, the

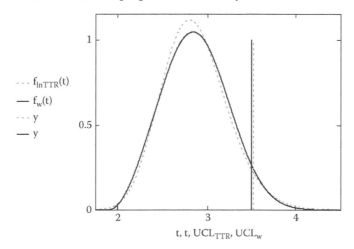

FIGURE B.7
Comparison between the log-normal and Weibull repair models.

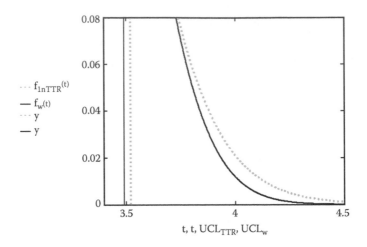

FIGURE B.8
Comparison of the UCL for the log-normal and Weibull repair models.

log-normal distribution will provide the error in the estimate of the MTTR and the upper confidence limit. The shape parameter for our example data is 2.88; close but not exactly equal to 2.26. The differences between the upper confidence limits are visible on the graph. The scale of the graph does not provide enough resolution to appreciate the differences.

The upper tails of the two distributions including the upper confidence limit locations are magnified in Figure B.8. It is evident from the figures that the log-normal distribution overstates the upper confidence limit for time to repair when the shape parameter is greater than 2.26 and understates the upper confidence limit when the shape parameter is less than 2.26.

Appendix C: Logistics Downtime Functions

Logistical downtime math models are victims of economic reality. All data cost money and few organizations are willing to invest in empirical data to characterize logistical downtime. Additionally, design organizations have no way to accurately estimate the logistical downtime experiences of the organizations that operate and maintain the systems they develop. Any estimate they use for logistical downtime to calculate the operational availability is a global estimate that rarely applies in the field. Management information systems used by organizations that operate and maintain systems do not track logistical downtime events.

The approach to fit logistics downtime math models by organizations that operate and maintain systems is an inexpensive approach that takes very little time and yields remarkably accurate estimates. Subject matter experts located in the organization understand the logistics downtime events that occur for systems that they operate and maintain and they have accurate, if not precise, knowledge of the duration of each event expressed as a range from minimum to maximum and a most likely value. A subjective probability distribution can fit this data to a triangular distribution as discussed in Chapter 3.

An example of findings for prerepair logistics downtime from a Delphi survey of subject matter experts is summarized in Table C.1. The range of the duration for prerepair logistics downtime is expressed as the minimum and maximum parameters of the triangular distribution. The most likely value is treated as the mode of the triangular distribution. As with time-to-repair experiments, the duration of logistics downtime events is expressed in minutes. The logistics downtime math models are fit by time units of hours.

The mean prerepair logistics downtime is the arithmetic average of the three parameters as expressed in the following equation.

$$\theta = \frac{T_{\min} + T_{\text{mode}} + T_{\max}}{3} = 9.893 \tag{C.1}$$

The prerepair logistics downtime math model is expressed as two intersecting lines that form a triangle that meets the condition of a probability distribution such that the area under the curve is equal to one. The intersection of the two lines occurs at the mode. Triangular distribution will fit skewed and symmetrical frequency distribution.

TABLE C.1

Prerepair Logistics Downtime Data

	T_{min}	T_{mode}	T_{max}
SME 1	69	90	194
SME 2	73	109	176
SME 3	66	122	176
SME 4	61	111	183
SME 5	72	95	184
	341	527	913
	5.68	8.78	15.22
			$\Lambda = 9.89$

$$
f(t) = \begin{cases}
0 & \text{if} \quad t < T_{min} \\[2mm]
\dfrac{2(t - T_{min})}{(T_{max} - T_{min})(T_{mode} - T_{min})} & \text{if} \quad T_{min} \le t \le T_{mode} \\[4mm]
\dfrac{2(T_{max} - t)}{(T_{max} - T_{min})(T_{max} - T_{mode})} & \text{if} \quad T_{mode} < t \le T_{max} \\[2mm]
0 & \text{if} \quad t > T_{max}
\end{cases}
\tag{C.2}
$$

The prerepair logistics math model is provided in Figure C.1.

The location of the mode is the line that extends to the intersection of the probability mass density function, and the location of the mean is shown by the dashed lines.

The median prerepair logistics downtime is expressed in the following equation.

$$
T_{med} = \begin{cases}
T_{min} + \sqrt{\dfrac{(T_{max} - T_{min})(T_{mode} - T_{min})}{2}} & \text{if} \quad \dfrac{T_{min} + T_{max}}{2} \le T_{mode} \\[5mm]
T_{max} - \sqrt{\dfrac{(T_{max} - T_{min})(T_{max} - T_{mode})}{2}} & \text{if} \quad \dfrac{T_{min} + T_{max}}{2} > T_{mode}
\end{cases}
\tag{C.3}
$$

A symmetrical triangular distribution meets the condition that the mean equals the median equals the mode. A positively skewed triangular distribution meets the condition that the mean is greater than the median, which is greater than the mode. A negatively skewed triangular distribution meets the condition that the mean is less than the median, which is less than the mode.

The cumulative prerepair logistics downtime math model is expressed in Equation C.4. The cumulative prerepair logistics downtime math model

$$T_{min} := 5.68 \qquad T_{mode} := 8.78 \qquad T_{max} := 15.22$$

$$\theta := \frac{T_{min} + T_{mode} + T_{max}}{3} = 9.893$$

$$f(t) := \begin{vmatrix} \dfrac{2 \cdot (t - T_{min})}{(T_{max} - T_{min}) \cdot (T_{mode} - T_{min})} & \text{if } T_{min} \le t \le T_{mode} \\[3ex] \dfrac{2 \cdot (T_{max} - t)}{(T_{max} - T_{min}) \cdot (T_{max} - T_{mode})} & \text{if } T_{mode} < t \le T_{max} \end{vmatrix}$$

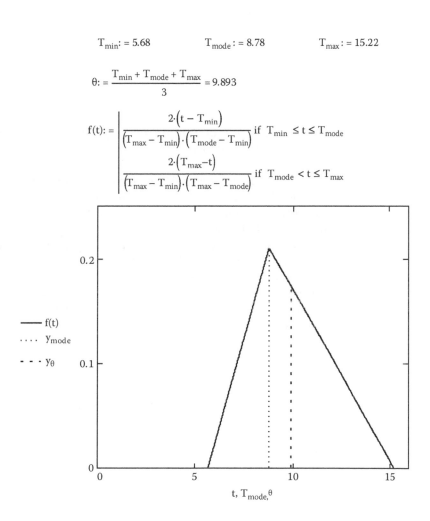

FIGURE C.1
Prerepair logistics math model.

enables the engineer to calculate the upper confidence limit for the prerepair logistics downtime.

$$F(t) = \begin{vmatrix} 0 & \text{if} & t < T_{min} \\ \dfrac{(t - T_{min})^2}{(T_{max} - T_{min})(T_{mode} - T_{min})} & \text{if} & T_{min} \le t \le T_{mode} \\ 1 - \dfrac{(-t)^2}{(T_{max} - T_{min})(T_{max} - T_{mode})} & \text{if} & T_{mode} < t \le T_{max} \\ 1 & \text{if} & t > T_{max} \end{vmatrix} \qquad (C.4)$$

The cumulative prerepair logistics downtime math model is provided in Figure C.2. The median represents the 50th percentile prerepair logistics downtime.

The upper confidence limit of the prerepair logistics downtime model is found by finding the value of the upper confidence limit for which the cumulative probability distribution is equal to $1 - \alpha$, or the confidence level.

A Mathcad routine is presented in Figure C.3. Alpha is stated as 0.05. The Boolean equivalence is entered for $1 - \alpha$ and the upper equation for the cumulative prerepair logistics downtime model. The Find(t) routine evaluates the Boolean equivalence and returns the two roots. The first root is greater than T_{max} and can the eliminated. The second root is selected and tested in the cumulative probability distribution. It is found to give the value 0.950, which is definitive for the upper confidence limit.

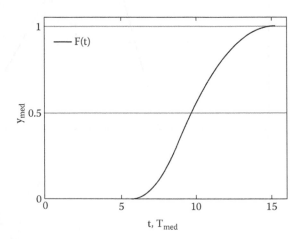

FIGURE C.2
Cumulative prerepair logistics downtime math model.

$$\alpha := 0.05$$

Given

$$1 - \alpha = 1 - \frac{\left(T_{max} - t\right)^2}{\left(T_{max} - T_{min}\right) \cdot \left(T_{max} - T_{mode}\right)}$$

Find (t) $\rightarrow$ (16.972677951022377642 13.467322048977622358)

UCL: = 13.467

FUCL = 0.950

FIGURE C.3

Mathcad solver routine for upper confidence limit.

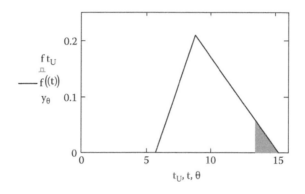

FIGURE C.4

Prerepair logistics downtime math model with upper confidence limit.

The prerepair logistics downtime math model with the shaded upper confidence limit is provided in Figure C.4.

The procedure for fitting the postrepair logistics downtime math model is identical.

The parameters of the prerepair logistics downtime math model are entered into the reliability simulation software program.

Appendix D: Engineering Economics Functions

Engineering economic analysis consists of cost estimation and evaluation. Cost estimation measures the magnitudes of cash flow events, sources and uses of cash, and when they occur in time. Cost evaluation compares two or more alternative cash flow events by calculating equivalent values. Appendix D provides the procedures for performing engineering economic analysis into sections

1. Most common analysis methods:
   a. Cost estimations for recurring, future, capital recovery, and sinking fund amounts
   b. Equivalent present and net present values for recurring and future amounts
   c. Equivalent uniform recurring amount for present and net present values

Engineering economic analysis textbooks provide equations for calculating recurring and future amounts and equivalent present and uniform recurring amounts. These equations were the predominant approach prior to the existence of spreadsheet software. Appendix D provides spreadsheet approaches exclusively, in the spirit of working smart rather than hard. Calculations are performed using spreadsheet economic functions unless economic equations are more appropriate.

2. Less common analysis methods include equivalent future values for present and recurring amounts. Equations and spreadsheet approaches are presented for less common analysis methods.

Appendix D presents procedures to estimate changes to cost events into the future. Cost estimation methods presented in Appendix D assume that determination of cost events is given, i.e., labor, materials, overhead to include calculations of the initial costs to the organization.

Engineering economic analysis applies end-of-period convention, which means that a cost that occurs in the first month of a project is allocated to the last day of the month. The determination of the time units for the analysis (weeks, months, quarters, or years) is a judgment call that is influenced by the demand for precision in the application of interest, escalation, and discount rates. Convenience also plays a part. A one-year project can conveniently be evaluated using weekly evaluation periods; a 20-year project would not. Similarly, a 20-year project can conveniently be evaluated using annual periods; a one-year project would not. Capital investment projects typically extend over 5 to 30 years and apply annual evaluation periods. Operating and sustainment engineering projects typically last less than five years and apply monthly, quarterly, and semiannual evaluation periods. The time units for engineering projects are typically influenced by the most prominent cost event frequency. Cost events that occur more frequently than the selected time units are presented as the sum that occurs for the selected time unit. For example, weekly labor expenses for monthly time units will be estimated for the monthly frequency, without regard to the impact of escalation factors that may occur within the selected time units. Loss of precision occurs when a significant disparity exists between the actual frequency of a cost event and the time unit. For example, weekly labor expenses that are estimated for semiannual frequencies may exclude provisions for a cost escalation that occurs within the six-month period. This is not a problem because the error across the evaluation alternatives will net out and not materially impact the comparative efficacy of the analysis.

This book uses the convention that cost events labeled as "amounts" (present amount, future amount, recurring amounts) represent cost estimations of actual cash flow to the organization, i.e., labor expenses, materials expenses, and overhead expenses. Growth of cash flows over time is estimated using escalation factors, g (referred to in engineering economics textbooks as a geometric gradient). One exception to the application of escalation factors is a linear gradient, G, which assumes that cash flows increase by constant steps over time.

Estimated interest expenses for capital investment, loans, bonds, etc., are cash flows that are estimated using interest rates, i. The interest rate that applies to an engineering project is the determining factor for the calculation of periodic capital recovery loan payments, term loan interest rate payments, and periodic sinking fund payments.

This book uses the convention that estimated equivalent present and recurring calculations for amounts are labeled as values (present value, future value, equivalent uniform recurring amount). Equivalent values are estimated using the organization's discount rate, its internal rate of return, sometimes referred to as the most attractive rate of return. The discount rate may also be the organization's cost of capital for debt and equity sources of cash. In the former case the discount rate is the minimum return on investment that an organization will allow to approve decision alternatives. In the latter case the discount rate represents the composite interest rate of all sources of cash and treats all cost events as repayments on debt.

It bears repeating that a distinction exists between amounts and values. Amounts represent the estimation of an actual cash flow event or a tax-deductible credit applied to an actual cash flow event. Values do not occur. For example, the present value of a series of future amounts represents an indifference criterion. An organization is indifferent between the choices of the actual cash flow of event versus possession of the equivalent present value based on the discount rate. Are you, the reader, indifferent between the promise of a dollar in one year versus having a dollar today? Assume that you can invest that dollar at 5% annual percentage rate (APR). The present value of a dollar in one year is $0.95. You are not indifferent between the two options; a dollar today has more value to you than a dollar a year from now. Assume instead that you are offered $1.05 a year from now versus a dollar today. The present value of the future amount is equal to $1.00, and you are indifferent between the two options. Let's make the analysis a little more complex. You have three options: $1.50 every month for two years, $3 every quarter for three years, or $10 at the end of 30 months. The magnitudes of the cost events vary for each option, the frequency of the cost events vary for each option, and the duration of the cost events vary for each option. The present value for each option is the indifference amount for each option. Had the durations of each option been equal, the present value provides the decision criterion: select the lowest cost or the highest revenue. Since the durations of each option are different, the equivalent recurring uniform amount calculated for the present value for each option provides the decision criterion.

Cost Estimation: Present Amount

Present amounts are estimations of all cash events that occur at the beginning of a project's evaluation period. Present amounts occur at $t = 0$ time units given a cash flow timeline that ranges from 0 to n time units. Examples of present amounts include initial investment in project startup, first lease payments on machinery and equipment, and license fees.

Cost Estimation: Future Amount

Future amounts are estimations of all unique cash events that occur at specific times during a project's evaluation period. Most but not all future amounts occur at the end of a project's evaluation period. Examples of future amounts include

Salvage Value at the End of the Useful Life of a Capital Asset

Salvage value is the estimated future amount that represents the residual value of a capital asset after a period of use. Salvage value is an amount

TABLE D.1

Salvage Value Calculation

PCA ($, at $t = 0$)	25,000
N (yr)	5
f_{SV}	0.85
F_{SV} ($, at $t = 5$)	11,093

required to calculate depreciation expense of the asset at the end of its useful life, or the disposal value at the end of the project period. Salvage value (F_{SV}) is calculated using an asset degradation rate (f_{SV}) as expressed in the following equation.

$$F_{SV} = f_{SV}^{N} P_{CA} \tag{D.1}$$

A spreadsheet solution to an example salvage value calculation is provided in Table D.1. The present value of a capital asset (P_{CA}) is estimated to be $25,000; the useful life of a capital asset is five years (N); and the engineer estimates the degradation rate to be 85% applied at the end of each year of use. Salvage value is estimated to be $11,093 at the end of five years.

Projected Balance of a Sinking Fund Investment

Sinking fund investments are used to provide an organization with the ability to pay a term loan at the end of its duration. Term loans incur interest payments over the duration of the loan with payment in full at the conclusion of the duration. A sinking fund is analogous to a savings account. The future amount of a sinking fund is equal to the present amount of the term loan. Organizations also use sinking funds to invest after-tax profits for future acquisitions or early retirement of capital recovery debt whose future amount is estimated based on the present amount and the expected cost escalation factor.

Unique Cash Events that Occur During the Evaluation Period

Unique cash events are specific expenses that occur singly at various times throughout the evaluation period. Salvage value and equipment overhaul are examples of unique cost events that occur at estimated points in time in the future.

Equivalent Present Value: Future Amount

The calculation of the present value of a future amount is expressed in the following equation.

$$PV = \frac{F}{(1+i)^n} \tag{D.2}$$

The interest rate in the equation (i) is the discount rate (r) expressed as the APR divided by the number of compounding periods (CP) per year (m), $i = r/m$. The exponent (n) of the quantity $(1 + i)$ is the total number of compounding periods for the project alternative and is equal to the product of the duration of the project in years (N) and the number of compounding periods per year, $n = mN$.

Microsoft Excel provides a formulas tab that includes financial functions. The present value function dialog box (PV) allows entry of the following four arguments:

- Rate: interest rate per compounding period, i
- Nper: total number of compounding periods for the project alternative, n
- Pmt: uniform recurring amount, A
- FV: future amount, F

The Excel PV function entered directly in a cell is PV(Rate, Nper, Pmt, FV). The PV function is capable of calculating the present value of a uniform recurring amount (Pmt) or future amount (FV), but it cannot do both simultaneously. Double commas following Nper provide a placeholder for the absent value of Pmt when calculating the present value of future amount, PV(Rate, Nper,, FV). Closing the present value function after the entry of Pmt tells Excel that there is no value for the future amount, PV(Rate, Nper, Pmt). An Excel example is presented in Table D.2.

Each future amount is an independent event and does not represent a recurring amount. Notice that the future amounts in years 2 and 4 describe expenses that are a use of cash and designated as negative (cash out), and the salvage value in year 5 is a positive value that represents the potential for

TABLE D.2

Present Value of Future Amount for $m = 1$

t (months)	F_t (\$)	PV_t (\$)	
0			
1			
2	−1,000.00	−841.68	−PV(0.09, 2,, −\$1,000)
3			
4	−3,000.00	−2,125.28	−PV(0.09, 4,, −\$3,000)
5	6,000.00	3,899.59	−PV(0.09, 5,, \$6,000)

Note: Given $F_2 = -\$1,000.00$; $F_4 = -\$3,000.00$; $F_{SV} = \$6,000.00/yr$; $r = 9\%$ APR; $m = 1$ CP/yr; and $i = 9\%$.

revenue (cash in). All uses of cash (operating expenses, interest payments) are negative, and all sources of cash (revenues, tax credits, salvage value) are positive. The reader should note that financial functions in Microsoft Excel reversed the sign, the present value of a negative cash flow amount will be positive and the present value of a positive cash flow amount will be negative. The engineer will logically determine that the present value of a negative amount should be negative, e.g., the sign convention should remain the same. A negative sign is entered preceding the financial function to adjust for this quirk.

Assume that this example uses a monthly time unit for the cash flow timeline. The future events occur in the 9th, 15th, and last (24th) time periods. The present value for each future amount is calculated for the interest rate per compounding period and the number of compounding periods, as illustrated in Table D.3.

Cost Estimation: Uniform Recurring Amounts

Uniform recurring amounts are the basic building blocks for the most prevalent cash events, periodic expenses for labor, materials, and overhead. Such periodic expenses may be estimated to be constant for short project durations. An example for cost estimation of uniform recurring amounts is presented in Figure D.1. The cost of an event is estimated to occur annually for five years at $100 per event.

Equivalent Present Value: Uniform Recurring Amount

Microsoft Excel, and any other spreadsheet program, is superior to the equation approach for calculating the present value of uniform recurring amounts. Two Excel approaches calculate the equivalent present value of a uniform recurring amount.

1. Excel financial function: PV(Rate, Nper, Pmt)

   The present value financial function, which evaluates a series of uniform recurring amounts, applies for a single calculation that does not involve any other amounts in the cash flow timeline. For example, given a uniform recurring amount ($A = \$100$) that occurs annually ($m = 1$) for five years ($N = 5$) at a discount rate ($r = 7.5\%$ APR), the present value (PV) is calculated in the following equation.

$$PV_A = \$404.59 = PV(0.075, 5, \$100)$$

   This approach has limited applicability.

TABLE D.3

Present Value of Future Amount for $m = 4$

t (months)	F_t ($)	PV_t ($)	
0			
1			
2			
3			
4			
5			
6			
7			
8			
9	−1,000	−966.87	−PV(0.00375, 9,, −$1,000)
10			
11			
12			
13			
14			
15	−3,000	−2,836.21	−PV(0.00375, 15,, −$3,000)
16			
17			
18			
19			
20			
21			
22			
23			
24	6,000	5,484.51	−PV(0.00375, 24,, $6,000)

Note: Given $F_9 = -\$1,000.00$; $F_{15} = -\$3,000.00$; $F_{SV} = \$6,000.00$; $N = 2$ yr; $r = 9\%$ APR; $m = 4$ CP/yr; $n = 24$ CP; and $i = 0.375\%$.

2. Excel financial function: PV(Rate, Nper,, FV)

Given the same example, we can treat each occurrence of the uniform recurring amount as a future amount that occurs in specific time periods, years 1, 2, … 5. The present value of each future amount is calculated for the year in which it occurs. The present value of the uniform recurring amounts is the sum of each year's present value as calculated in Table D.4.

This approach has universal applicability. Consider a cash flow timeline that sums the net cash flow that occurs in each month over the five-year duration of a project, 60 rows of recurring events (not necessarily uniform). Spreadsheets allow entering the present value

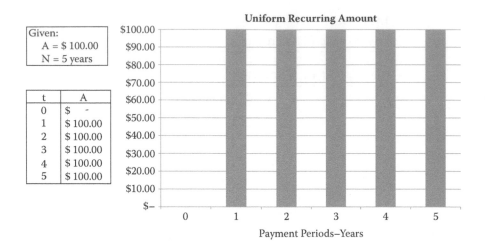

FIGURE D.1
Cost estimation - uniform recurring amounts – m = 1.

TABLE D.4

Present Value Uniform Recurring Amount $m = 1$

t (months)	A ($)	PV_t ($)	
0			
1	100.00	93.02	= PV(0.075, 1, $100)
2	100.00	86.53	= PV(0.075, 2, $100)
3	100.00	80.50	= PV(0.075, 3, $100)
4	100.00	74.88	= PV(0.075, 4, $100)
5	100.00	69.66	= PV(0.075, 5, $100)
	$PV_A = \Sigma PV_t$	404.59	

Note: Given A = $100.00; N = 5 yr; r = 7.50% APR;

function wants for the first time period and dragging the equation down the column to calculate the present value of each of the remaining 59 rows. The net present value (NPV) is the sum of the present values per period. Note that the present values of the uniform recurring amount are equal for both approaches.

Recurring Amounts with Linear Gradient

The majority of recurring amounts is not uniform. It is rare to encounter a uniform recurring amount in an engineering project except for very short

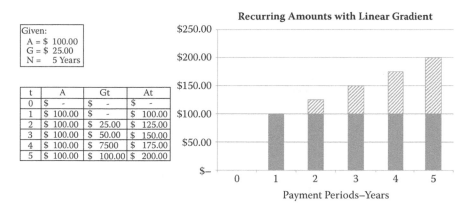

Given:
A = $ 100.00
G = $ 25.00
N = 5 Years

t	A	Gt	At
0	$ -	$ -	$ -
1	$ 100.00	$ -	$ 100.00
2	$ 100.00	$ 25.00	$ 125.00
3	$ 100.00	$ 50.00	$ 150.00
4	$ 100.00	$ 7500	$ 175.00
5	$ 100.00	$ 100.00	$ 200.00

FIGURE D.2
Cost estimation recurring amount with linear gradient m = 1.

duration projects. Occasionally, a recurring amount will increase over time by a constant increment (*G*) linear gradient. The application of the linear gradient occurs after the first amount (*A* + *G*) and increases each year by the value of linear gradient. The recurring amount is expressed as A_t, the value of the recurring amount in period *t*, as illustrated in Figure D.2.

Equivalent Present Value: Recurring Amounts with Linear Gradient

The present value of recurring amounts with a linear gradient required a calculation of the present value of the uniform recurring amount plus the present value of the gradient prior to availability of spreadsheet programs. The two equations are inconvenient for inclusion in a spreadsheet cell and are not necessary. The approach to treat each recurring amount as a future amount over the duration of the project, as shown in the preceding example for uniform recurring amounts, is equally precise as the equation approach and far more efficient. The present value (PV_t), of the amount of A_t for any value of *t* is calculated by the excel PV function as

$$PV_t = PV(\text{Rate}, \text{Nper}, \_, FV_t)$$

where Rate is the discount rate, divided by the number of compounding periods per year (*m*), Nper is the value of *t* in units of the number of compounding periods per year, and FV_t is the recurring amount that occurs at *t*.

This approach to calculate the present value for recurring amounts with a linear gradient given a discount rate of 7.5% APR is illustrated in Table D.5.

The frequency of the application of the linear gradient often differs from the frequency of the recurring amount. The initial value of the recurring amount may be uniform for a specified period before the gradient is applied; subsequently the gradient is applied at its designated frequency, resulting

TABLE D.5

Present Value Recurring Amount
with Linear Gradient $m = 1$

t (months)	A_t (\$)	PV_t (\$)
0		
1	100	93.02
2	125	108.17
3	150	120.74
4	175	131.04
5	200.0	139.31
	$PV_A = \Sigma PV_t$	592.29

Note: Given $A = \$100.00/$month; $N = 5$ yr; $r = 7.50\%$ APR; $G = \$25/$yr; $m = 1$ CP/yr; $i = 0.075$ r/CP; and $n = 5$ CP.

in a series of increasing uniform recurring amounts. Table D.6 demonstrates a recurring amount that occurs monthly with a linear gradient that is applied quarterly.

The cash flow timeline for the recurring amount with linear gradients and different respective frequencies is provided in Figure D.3.

Engineering economic analysis lacks an equation for this scenario. Using the approach that treats each recurring amount as a future amount is the only reasonable approach to finding the present value. The calculation for the present value of the recurring amounts is provided in Table D.7.

Recurring Amounts with Geometric Gradient

Recurring amounts with geometric gradient are the most common recurring amount scenario. Most operating costs, labor, materials, and overhead have an initial value at the beginning of the evaluation period and increase by a cost escalator, g, geometric gradient. Labor cost escalators include cost of living increases based on the inflation rate and negotiated contract provisions. Materials cost escalators are provided by the Department of Commerce for all categories of materials and have as many cost escalators as there are categories of materials. Overhead cost escalators are calculated by the organization and are more complex because they include fixed costs that do not vary over time (cost allocations of floor space determined by the financing of the faculty) with the increasing costs of labor and materials. For example, a recurring cost has an initial value of $100, a cost escalator of 5% APR, and the

TABLE D.6

Cost Estimation Recurring Amount with $m = 12$ and Linear Gradient with $m = 4$

t (months)	A ($)	G_t ($)	A_t ($)
0			
1	100	0	100
2	100	0	100
3	100	0	100
4	100	25	125
5	100	25	125
6	100	25	125
7	100	50	150
8	100	50	150
9	100	50	150
10	100	75	175
11	100	75	175
12	100	75	175
13	100	100	200
14	100	100	200
15	100	100	200
16	100	125	225
17	100	125	225
18	100	125	225
19	100	150	250
20	100	150	250
21	100	150	250
22	100	175	275
23	100	175	275
24	100	175	275

Note: Given $A = \$100.00$/month; $N = 2$ yr; $r = 7.50\%$ APR; $G = \$25$/qtr; $m = 12$ CP/yr; $i = 0.0063$ r/CP; and $n = 24$ CP.

duration of five years. The initial value of recurring costs (T) is $100. The cost escalator is applied at the end of the second period as shown in Figure D.4.

Equivalent Present Value: Recurring Amounts with Geometric Gradient

Engineering economics textbooks that present equations to calculate the equivalent present value for recurring amounts of geometric gradient recognize two conditions: (1) the cost escalation factor, g, is different than the discount rate, r; (2) cost escalation factor equals the discount rate. Each condition

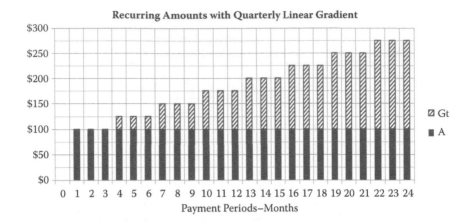

FIGURE D.3
Cost estimation recurring amount with linear gradient m = 4.

has its own equation. By now the reader has detected a pattern in the application of spreadsheets to perform engineering economics analysis. The pattern is a two-step procedure: (1) estimate the recurring amount, (2) calculate the present value for each period and add them up. This procedure is illustrated in Table D.8 for the equivalent present value of recurring amounts with a geometric gradient.

The frequency of the recurring amount will often differ from the frequency of the geometric gradient, just as with the linear gradient. In the following example, initial value of the linear gradient is $100. The frequency of the application of the geometric gradient is quarterly, every three months. The combined estimation of the recurring amounts and the corresponding present value is provided in Table D.9.

The cash flow timeline for the recurring amounts with geometric gradient is presented in Figure D.5.

Capital Recovery

Capital recovery is a special case of the uniform recurring amount. Capital recovery describes a loan instrument, not unlike a consumer car payment or house mortgage, in which the organization makes equal periodic payments to pay down the loan. Each payment consists of interest on the outstanding balance of the loan and payment on the principal to reduce the outstanding balance of the loan. The engineer and manager must break down the uniform recurring amount to identify the two individual cash flows. The interest portion of the uniform recurring amount is a tax-deductible expense

TABLE D.7

Present Value of Recurring
Amount with $m = 12$ and Linear
Gradient with $m = 4$

t (months)	A_t ($)	PV_t ($)
0		
1	100.00	99.38
2	100.00	98.76
3	100.00	98.15
4	125.00	121.92
5	125.00	121.17
6	125.00	120.41
7	150.00	143.60
8	150.00	142.71
9	150.00	141.82
10	175.00	164.43
11	175.00	163.41
12	175.00	162.39
13	200.00	184.44
14	200.00	183.29
15	200.00	182.16
16	225.00	203.65
17	225.00	202.39
18	225.00	201.13
19	250.00	222.09
20	250.00	220.71
21	250.00	219.34
22	275.00	239.77
23	275.00	238.29
24	275.00	236.81
	$PV_A = \Sigma PV_t =$	4,112.21

Note: Given $N = 2$ yr; $r = 7.50\%$ APR;
$m = 12$ CP/yr; $i = 0.00625$
r/CP; and $n = 24$ CP.

that generates a tax credit for the organization. The payment on principal is not a tax-deductible expense, although it is an actual cash transaction.

The uniform recurring amount for a capital recovery loan uses the present amount of the loan to calculate the uniform recurring amount, $A_{CR} =$ Pmt(Rate, Nper, PV). "Rate" is equal to the loan interest expressed as an annual percentage rate divided by the number of loan payments per year. "Nper" is equal to the product of the duration of the loan in years and the number of loan payments per year. The Microsoft Excel finance function

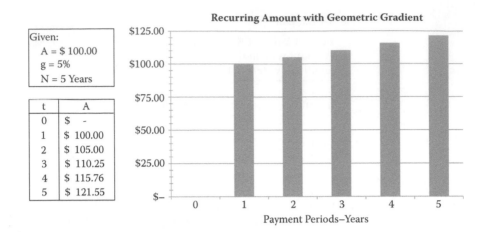

t	A
0	$ -
1	$ 100.00
2	$ 105.00
3	$ 110.25
4	$ 115.76
5	$ 121.55

Given:
A = $ 100.00
g = 5%
N = 5 Years

FIGURE D.4
Cost estimation recurring amount with geometric gradient m = 1.

TABLE D.8

Present Value Recurring
Amount with Geometric
Gradient $m = 1$

t (months)	A_t ($)	PV_t ($)
0		
1	100.00	93.02
2	105.00	90.86
3	110.25	88.75
4	115.76	86.68
5	121.55	84.67
	$PV_A = \Sigma PV_t$	443.98

Note: Given A = $100.00/yr; $r = 7.50\%$ APR; $g = 5\%$ APR; $N = 5$ yr; $m = 1$ CP/yr; $i = 0.075\,r/CP$; and $n = 5$ CP.

for a uniform recurring amount, A, is Pmt(Rate, Nper, PV, FV) and can be used to find the uniform recurring amount for either a present amount for a future amount.

The following example expresses the present amount of a loan (P) and the amount of $1,000. The loan interest rate (r) is 5% APR. The loan duration is five years with annual payments. The uniform recurring amount (A_{CR}) is $230.97. The first payment is comprised of the interest on the current loan balance (Pmt_{int} = $50.00), equal to 5% times $1000, and the payment on principal (Pmt = $180.97), equal to $230.97 − $50.00. The current loan balance is

TABLE D.9

Cost Estimation and Present Value Recurring Amount with Geometric Gradient $m = 4$

t (months)	A_t ($)	PV_t ($)
0		
1	100.00	99.38
2	100.00	98.76
3	100.00	98.15
4	105.00	102.42
5	105.00	101.78
6	105.00	101.15
7	110.25	105.54
8	110.25	104.89
9	110.25	104.24
10	115.76	108.77
11	115.76	108.09
12	115.76	107.42
13	121.55	112.09
14	121.55	111.40
15	121.55	110.71
16	127.63	115.52
17	127.63	114.80
18	127.63	114.09
19	134.01	119.05
20	134.01	118.31
21	134.01	117.57
22	140.71	122.69
23	140.71	121.92
24	140.71	121.17
$PV_A = \Sigma PV_t$		2,639.90

Note: Given A = $100.00; N = 5 yr; g = 5%/qtr; r = 7.50% APR; m = 12 CP/yr; i = 0.00625 r/CP; and n = 60 CP.

reduced to $819.03. The second payment is comprised of the reduced interest on the new loan balance and the increased payment on principal, as illustrated in Figure D.6. The last payment reduces the loan balance to $0.00. The project cash flow timeline will include the interest payments for the period to calculate the organization's tax credit. The net cash flow for the period will be equal to the uniform recurring amount plus the tax credit.

Taxes represent a very complex set of rules that can only be understood by the accounting department. Engineers and managers are encouraged to

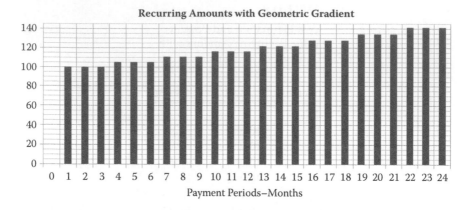

FIGURE D.5
Cost estimation recurring amount with geometric gradient m = 4.

Given:
P = $ 1,000
P = 5%
N = 5 years

t	A	Balance	PMT_{int}	PMT_{Prin}
0		$ 1,000.00		
1	$230.97	$ 819.03	$ 50.00	$ 180.97
2	$230.97	$ 629.00	$ 40.95	$ 190.02
3	$230.97	$ 429.48	$ 31.45	$ 199.52
4	$230.97	$ 219.98	$ 21.47	$ 209.50
5	$230.97	$ 0.00	$ 11.00	$ 219.98

FIGURE D.6
Capital recovery m = 1.

include the organization's accountants in performing reliability-based life-cycle economic analyses to correctly apply the tax credits to the cost events.

Sinking Fund

Sinking funds are another special case of the uniform recurring amount. Sinking funds describe an investment made by the organization to pay a term loan instrument. Sinking fund investments are uniform recurring amounts (Pmt_{SF}) that are based on the future amount required to pay the term loan. The organization pays simple interest on the term loan that becomes a uniform recurring amount in the cash flow timeline. The Microsoft Excel calculation for the sinking fund uniform recurring amount is A_{SF} = Pmt(Rate, Nper,, FV), as shown in Figure D.7. The accrued balance of the sinking fund

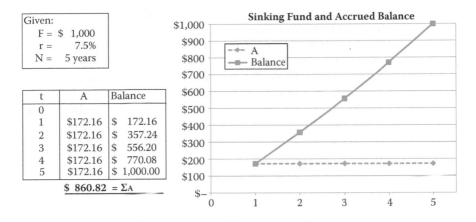

Given:

F =	$ 1,000
r =	7.5%
N =	5 years

t	A	Balance
0		
1	$172.16	$ 172.16
2	$172.16	$ 357.24
3	$172.16	$ 556.20
4	$172.16	$ 770.08
5	$172.16	$ 1,000.00

$ 860.82 $= \Sigma$A

FIGURE D.7
Sinking fund m = 1.

grows to equal the balance of the term loan from the compound interest. Although sinking funds are actual cash flows in the cash flow timeline and are included in the calculation of the net cash flow per payment period, sinking funds are not tax-deductible expenses. Sinking fund investment comes from an organization's after-tax income.

Net Present Value and Equivalent Uniform Recurring Amount

Net present value (NPV) and the equivalent uniform recurring amount (A_{Equiv}) are the two criteria that are used to evaluate and select engineering economic alternatives.

Net Present Value

The net present value is calculated from a cash flow timeline that includes the following:

- All present amounts that occur at the initiation of the project, P
- All future amounts that occur during the project, F
- All recurring amounts that occurred during the project, A
- Internal worksheet that calculates the interest portion of capital recovery loans
- Internal worksheet that calculates the depreciation expense for capital assets
- Internal worksheet that calculates the tax credit for each time period
- Net cash flow per time period

There are two options to calculate the net present value in Microsoft Excel: (1) calculate the present value of each time period as discussed in the preceding examples, and (2) the net present value financial function, NPV. The net present value function is expressed in a cell as NPV = NPV(Rate, $NCF_{t = 1}$:$NCF_{t = n}$), where the rate is the discount rate per payment period, and $NCF_{t = 1}$:$NCF_{t = n}$ is the selection of all of the values in the net cash flow column from $t = 1$ to $t = n$. The net present value function provides the net present value of all net cash flows except the present amounts that exist in the period $t = 0$. The net present value of the project adds the present amounts to the NPV function.

Equivalent Uniform Recurring Amount

The equivalent uniform recurring amount (A_{Equiv}) is calculated for the NPV. The Microsoft Excel financial function is identical to that used for any uniform recurring amount calculation for a present or a future amount, A_{Equiv} = Pmt(Rate, Nper, PV), where PV is equal to NPV.

An example cash flow timeline is presented in Table D.10. The net cash flow column (NCF_t) is the sum of all cash flow events for each period, t. Both approaches for calculating the net present value are provided in the cash flow timeline: (1) NPV = ΣPV_t, and (2) NPV = NPV(i, $NCF_{t = 1}$:$NCF_{t = n}$) + NCF_t = 0 ($NCF_{t = 0}$ = $PV_{t = 0}$).

$$NPV = NPV(i, NCF_{t = 1}:NCF_{t = n}) + NCF_{t = 0}$$

$$A_{equiv} = Pmt(i, n, NPV)$$

TABLE D.10

Cash Flow Timeline NPV of Equivalent Uniform Recurring Amount

t (months)	P ($)	A ($)	A_G ($)	A_g ($)	F ($)	NCF_t ($)	PV_t ($)
0	1,000					1,000.00	1,000.00
1		100	100	100.00		300.00	279.07
2		100	125	110.00	500	835.00	722.55
3		100	150	121.00		371.00	298.64
4		100	175	133.10	750	1,158.10	867.19
5		100	200	146.41	1,000	1,446.41	1,007.51
						4,174.96	4,174.96

$$NPV = NPV(i, NCF_{t = 1}:NCF_{t = n}) + NCF_{t = 0}$$
$$NPV = \Sigma PV_t$$
$$A_{equiv} = Pmt(i, n, NPV) \quad 1,031.90$$

Note: Given r_{IRR} = 7.50% APR; G = \$25.00; g = 10%; N = 5; m = 1; i = 7.50%; and n = 5.

Less Common Engineering Economics Functions

Engineering economic analysis typically starts with cost estimation for cash events and places them in the time period in which they occur followed by calculation of the their respective equivalent present and recurring uniform values. Engineering economic financial functions include calculating equivalent future values for present and recurring amounts. Equivalent future values are not employed in evaluation of life-cycle economic analysis but are presented here for the reader's consideration.

Future Value (*F*) Given Present Amount (*P*)

Inputs: present amount, *P*; discount rate, *r*; duration, *N*; and compounding periods per year, *m*.

Approach: Find interest rate per compounding period, $i = g/m$, and total compounding periods, $n = mN$.

Equation

$$F = P(1+i)^n \qquad\qquad (D.3)$$

Excel Function

$$F = \text{FV(Rate, Nper, \_, PV)} \sim \text{FV}(i, n, \_, P)$$

Example 1: Annual Compounding Periods

Given

- $P = 100$
- $r = 5\%$
- $N = 5$
- $m = 1$

Solve for

- $i = r/m = 5\%/1 = 0.05$
- $n = mN = (1)(5) = 5$

Find: *F*
Equation

$$F = 100(1.05)^5 = 127.63$$

Excel Function

$$F = \$127.63 = \text{FV}(0.05, 5,, 100)$$

Example 2: Monthly Compounding Periods

Given

- $P = 100$
- $r = 5\%$
- $N = 5$
- $m = 12$

Solve for

- $i = r/m = 5\%/12 = 0.00417$
- $n = mN = (12)(5) = 60$

Find F:
Equation

$$F = 100(1.00417)^{60} = 128.36$$

Excel Function

$$F = \$128.36 = FV(0.00417, 60,, 100)$$

Future Value in Time (F_t) Given Present Amount (P)

Inputs: present value, P; discount rate, r; duration, N; and compounding periods per year, m.
   Approach: Find interest rate per compounding period, $i = g/m$.
   Equation

$$F_t = P(1+i)^t \tag{D.4}$$

Excel Function

$$F_t = FV(Rate, Nper, \_, PV) \sim FV(i, t, \_, P)$$

Example 1: Annual Compounding Periods

Given

- $P = 100$
- $r = 5\%$
- $N = 5$
- $m = 1$

TABLE D.11

Future Value of Present Amount $m = 1$ by Equation

t (months)	P (\$)	F_t	
		$= P(1 + i)t$ (\$)	$= F_t - 1(1 + i)$ (\$)
0	100.00		$F_t = 0 = P$
1		105.00	105.00
2		110.25	110.25
3		115.76	115.76
4		121.55	121.55
5		127.63	127.63

TABLE D.12

Future Value of Present Amount $m = 1$ by Excel Financial Function

t (months)	P (\$)	F_t (\$)
0	100.00	
1		105.00
2		110.25
3		115.76
4		121.55
5		127.63

Solve for

- $i = r/m = 5\%/1 = 0.05$
- $t = 1, 2, \ldots mN = 1, 2, \ldots(1)(5) = 1, 2, \ldots 5$

Find F_t

Equation in Excel Spreadsheet (Table D.11)

Excel Function (Table D.12)

Future Value for Recurring Amounts

Spreadsheet software like Excel can calculate future values for uniform recurring amounts (A) only. Future value functions are not available in spreadsheets for increasing recurring amounts (linear and geometric), gradients, and recurring amounts that are uniform for a period of time and then increase periodically. Equations are available to calculate future values for increasing recurring amounts. The equation for the future value for a linear increasing recurring amount follows:

Let

A = uniform recurring amount

G = linear gradient

i = interest rate per compounding periods

n = total compounding periods

F = future value for a linear increasing recurring amount = $F_A + F_G$, where

F_A = future value for a uniform recurring amount

F_G = future value for a linear gradient

Equations

$$F_A = A\frac{(1+i)^n - 1}{i} \tag{D.5}$$

$$F_G = \left(\frac{G}{i}\right)\left[\frac{(1+i)^n - 1}{i} - n\right] \tag{D.6}$$

$$F = F_A + F_G \tag{D.7}$$

The equation for the future value for a geometric increasing recurring amount follows:

Let

A_1 = initial amount of the recurring amounts

g = geometric gradient: the rate at which the recurring amount increases each compounding period

i = interest rate per compounding periods

n = total compounding periods

F = future value for a linear increasing recurring amount under the following conditions:

$r \ne i$

$r = i$

Equation:

For $r \neq i$

$$F = A_1 \left[\frac{(1+i)^n - (1+g)^n}{i-g} \right] \tag{D.8}$$

For $r = i$

$$F = nA_1 (1+i)^{n-1} \tag{D.9}$$

It bears repeating that future value equations for linear and geometric increasing recurring amounts only apply when the frequency of the gradients is identical to the frequency of the compounding periods. For example, the equations do not apply when the gradient frequency is annual and the compounding frequency is monthly.

Excel Function: $F = \text{FV}(\text{Rate, Nper, Pmt}) = \text{FV}(i, n, A)$

Example 1: Annual Compounding Periods

Given

- $A = 100$
- $r = 5\%$
- $N = 5$
- $m = 1$

Solve for

- $i = r/m = 5\%/1 = 0.05$
- $t = 1, 2, \ldots mN = 1, 2, \ldots(1)(5) = 1, 2, \ldots 5$

Excel approach (Table D.13)

$$\text{FV} = \text{FV}(\text{Nper, Rate, Pmt}) = \text{FV}(0.05, 5, \$100) = \$552.56$$

TABLE D.13

Future Value for Uniform Recurring
Amount: $m = 1$

Find F	
F (\$)	552.56

Note: Given $A = \$100.00$; $r = 5\%$ APR; $N = 5$ yr; $m = 1$ CP/yr; $i = 0.05/\text{CP}$; and $n = 5$ CP.

TABLE D.14

Future Value for Uniform Recurring
Amount: $m = 12$

Find F	
F ($)	566.72

Note: Given $A = \$8.33$; $r = 5\%$ APR; $N = 5$ yr; $m = 12$ CP/yr; $i = 0.00417$/CP; and $n = 60$ CP.

Example 2: Monthly Compounding Periods

Given

- $A = 100/12$
- $r = 5\%$
- $N = 5$
- $m = 1$

Solve for

- $i = r/m = 5\%/1 = 0.05$
- $t = 1, 2, \ldots mN = 1, 2, \ldots(1)(5) = 1, 2, \ldots 5$

Excel Function (Table D.14)

$$F = FV(0.00417, 60, \$8.33) = \$566.72$$

Disclaimer: the author has no affiliation with Microsoft Corporation. Microsoft Office is not the only software program available. The author has encountered many other spreadsheet products that have been used by students in his engineering economics courses. All of those spreadsheet products replicated the Microsoft financial functions and performed equally well.

References

Abernathy, R.B. 2006. *The New Weibull Handbook,* 5th Ed. Robert B. Abernathy, North Palm Beach, FL.

Ahmad, R., & Kamaruddin, S. 2012. "An Overview of Time-Based and Condition-Based Maintenance in Industrial Application." *Computers & Industrial Engineering,* 63(1), 135–149.

AMCP 706-196. 1976. *Engineering Design Handbook: Development Guide for Reliability.* HQ US Army Materiel Command.

Azarkhail, M., & M. Modarres. 2007. "Markov Chain Simulation for Estimating Accelerated Life Model Parameters." Proceedings of the Reliability and Maintainability Symposium, Orlando, FL.

Bayoumi, A., N. Goodman, R. Shah, T. Roebuck, A. Jarvie, L. Eisner, L. Grant, & J. Keller. 2008. "Conditioned-Based Maintenance at USC—Part III: Aircraft Components Mapping and Testing for CBM." Proceedings of the American Helicopter Society Specialists Meeting on Condition Based Maintenance, Huntsville, AL.

Bayoumi, A., N. Goodman, R. Shah, T. Roebuck, A. Jarvie, L. Eisner, L. Grant, & J. Keller. 2008. "Conditioned- Based Maintenance at USC—Part I: Integration of Maintenance Management Systems and Health Monitoring Systems through Historical Data Investigation." Proceedings of the American Helicopter Society Specialists Meeting on Condition Based Maintenance, Huntsville, AL.

Bazargan, M., & R.N. McGrath. 2003. "Discrete Event Simulation to Improve Aircraft Availability and Maintainability." Proceedings of the Reliability and Maintainability Symposium, Tampa, FL.

Bazovsky, I. 1961. *Reliability Theory and Practice.* Prentice-Hall, Upper Saddle River, NJ.

Birolini, A. 1994. *Reliability Engineering: Theory and Practice,* 4th Ed. Swiss Federal Institute of Technology, Zurich.

Black, P.H., & O.E. Adams 1968. *Machine Design,* 3rd Ed. McGraw-Hill Book Co., New York, NY.

Blanchard, B.S. 2004. *Logistics Engineering and Management,* 6th Ed. Prentice-Hall, Upper Saddle River, NJ.

Brall, A., W. Hagen, & H. Tran. 2007. "Reliability Block Diagram Modeling—Comparisons of Three Software Packages." Proceedings of the Reliability and Maintainability Symposium, Orlando, FL.

Briand, D., & J.E. Campbell. 2007. "Real Time Consequence Engine." Proceedings of the Reliability and Maintainability Symposium, Orlando, FL.

Carter, C.M., & A.W. Malerich. 2007. "The Exponential Repair Assumption: Practical Impacts." Proceedings of the Reliability and Maintainability Symposium, Orlando, FL.

Collins, J.A. 1993. *Failure of Materials in Design Analysis Prediction Prevention,* 2nd Ed. John Wiley & Sons, New York, NY.

Componation, P.J., P.W. Jemison, W.R. Wessels, & S. Gholston. Spring 2011. "Exploring Alternate Strategies for Highly Accelerated Life Testing." The Journal of Reliability, Maintainability, Supportability in Systems Engineering, Frederick, MD.

Condra, L.W. 1993. *Reliability Improvement with Design of Experiments.* Marcel Dekker, Inc., New York, NY.

Cook, J. 2009. "System of Systems Reliability for Multi-State Systems." Proceedings of the Reliability and Maintainability Symposium, Ft Worth, TX.

Department of Defense. 2008. *CBM DoD Guidebook.* DoD, Washington, DC.

Distefano, S., & A. Puliafito. 2007. "Dynamic Reliability Block Diagrams vs Dynamic Fault Trees." Proceedings of the Reliability and Maintainability Symposium, Orlando, FL.

Dodson, B. 1994. *Weibull Analysis.* ASQ Quality Press, Milwaukee, WI.

Dovich, R.A. 1990. *Reliability Statistics.* ASQ Quality Press, Milwaukee, WI.

Draper, N.R., & H. Smith 1981. *Applied Regression Analysis,* 2nd Ed. John Wiley & Sons, New York, NY.

Ebeling, C. 2000. *An Introduction to Reliability and Maintainability Engineering.* McGraw-Hill, New York, NY.

Farquharson, J.A., & J.L. McDuffee. 2003. "Using Quantitative Analysis to Make Risk-Based Decisions." Proceedings of the Reliability and Maintainability Symposium, Tampa, FL.

Farrington, S.E., Sillivant, D., & Sautter, C. 2011. "Reliability Centered Maintenance as Applied to Wind Turbine Power Plants." In *ASME International Mechanical Engineering Congress and Exposition 2011.* Denver, CO, pp. 1–5.

Goel, H., J. Grievink, P. Herder, & M. Weijnen. 2003. "Optimal Reliability Design of Process Systems at the Conceptual Stage of Design." Proceedings of the Reliability and Maintainability Symposium, Tampa, FL.

Hartog, J.P. 1984. *Mechanical Vibrations,* 4th Ed. Dover Publications, Inc., New York, NY.

Hauge, B.S., & B.A. Mercier. 2003. "Reliability Centered Maintenance Maturity Level Roadmap." Proceedings of the Reliability and Maintainability Symposium, Tampa, FL.

Hicks, C.R. 1993. *Fundamental Concepts in the Design of Experiments,* 4th Ed. Saunders College Publishing, New York, NY.

Ireson, W.G., C.F. Coombs, & R.Y. Moss 1996. *Handbook of Reliability Engineering and Management,* 2nd Ed. McGraw-Hill Book Co., New York, NY.

Kapur, K.C., & L.R. Lamberson. 1977. *Reliability in Engineering Design.* John Wiley & Sons, New York, NY.

Kececioglu, D. 2002. *Reliability Engineering Handbook,* Vol. 2 Revised. DEStech Publications, Inc.

Kobbacy, K.A.H., & Murthy, D.N.P. eds. 2008. *Complex System Maintenance Handbook,* 2008 Edition. Springer.

Krasich, M. 2003. "Accelerated Testing for Demonstration of Product Lifetime Reliability." Proceedings of the Reliability and Maintainability Symposium, Tampa, FL.

Krasich, M. 2007. "Realistic Reliability Requirements for the Stresses in Use." Proceedings of the Reliability and Maintainability Symposium, Orlando, FL.

Krasich, M. 2009. "How to Estimate and Use MTTF/MTBF: Would the Real MTBF Please Stand Up?" Proceedings of the Reliability and Maintainability Symposium, Ft Worth, TX.

Krishnamoorthi, K.S. 1992. *Reliability Methods for Engineers.* ASQC Quality Press, Milwaukee, WI.

Lambeck, R.P. 1983. *Hydraulic Pumps and Motors: Selection and Application for Hydraulic Power Control Systems.* Marcel Dekker, Inc., New York, NY.

Lanza, G., P. Werner, & S. Niggeschmidt. 2009. "Adapted Reliability Prediction by Integrating Mechanical Load Impacts." Proceedings of the Reliability and Maintainability Symposium, Ft Worth, TX.

Lanza, G., P. Werner, & S. Niggeschmidt. 2009. "Behavior of Dynamic Preventive Maintenance Optimization for Machine Tools." Proceedings of the Reliability and Maintainability Symposium, Ft Worth, TX.

Lapin, L.L. 1998. *Probability and Statistics for Modern Engineering,* 2nd Ed. Waveland Press, Inc., Prospect Heights, IL.

Leemis, Lawrence 1995. *Reliability Probabilistic Models and Statistical Methods.* Prentice-Hall, Upper Saddle River, NJ.

Lefebvre, Y. 2003. "Using Equivalent Failure Rates to Assess the Unavailability of an Ageing System." Proceedings of the Reliability and Maintainability Symposium, Tampa, FL.

Liddown, M., & G. Parlier. 2008. "Connecting CBM to the Supply Chain: Condition Based Maintenance Data for Improved Inventory Management and Increased Readiness." Proceedings of the American Helicopter Society Specialists Meeting on Condition Based Maintenance, Huntsville, AL.

Liu, Y., H-Z. Huang, & M.J. Zuo. 2009. "Optimal Selective Maintenance for Multi-State Systems under Imperfect Maintenance." Proceedings of the Reliability and Maintainability Symposium, Ft Worth, TX.

Luo, M., & T. Jiang. 2009. "Step Stress Accelerated Life Testing Data Analysis for Repairable System Using Proportional Intensity Model." Proceedings of the Reliability and Maintainability Symposium, Ft Worth, TX.

Mannhart, A., A. Bilgic, & B. Bertsche. 2007. "Modeling Expert Judgment for Reliability Prediction—Comparison of Methods." Proceedings of the Reliability and Maintainability Symposium, Orlando, FL.

MIL-HDBK-472. 1966. "Maintainability Prediction." Military Standardization Handbook. DoD, Washington, DC.

MIL-STD-471. 1973. "Maintainability Verification/Demonstration/Evaluation." Military Standardization Handbook. DoD, Washington, DC.

MIL-STD-785. 1980. "Reliability Program for Systems & Equipment Development & Production." Military Standardization Handbook. DoD, Washington, DC.

MIL-STD-1629. 1980. "Procedures for Performing a Failure Mode, Effects & Criticality Analysis." Military Standardization Handbook. DoD, Washington, DC.

MIL-STD-756. 1981. "Reliability Modeling & Prediction." Military Standardization Handbook. DoD, Washington, DC.

MIL-STD-2155. 1985. "Failure Reporting, Analysis & Corrective Action System (FRACAS)." Military Standardization Handbook. DoD, Washington, DC.

MIL-STD-781. 1986. "Reliability Testing for Engineering Development, Qualification & Production." Military Standardization Handbook. DoD, Washington, DC.

MIL-STD-470. 1989. Maintainability Program for Systems & Equipment. Military Standardization Handbook. DoD, Washington, DC.

MIL-STD-690. 1993. "Failure Rate Sampling Plans & Procedures." Military Standardization Handbook. DoD, Washington, DC.

MIL-HDBK-781. 1996. "Reliability Test Methods, Plans and Environments for Engineering Development, Qualification & Production." Military Standardization Handbook. DoD, Washington, DC.

MIL-STD-810. 2008. "Environmental Test Methods and Engineering Guidelines." Military Standardization Handbook. DoD, Washington, DC.

Misra, R.B., & B.M. Vyas. 2003. "Cost Effective Accelerated Testing." Proceedings of the Reliability and Maintainability Symposium, Tampa, FL.

Montgomery, D.C., G.C. Runger, & N.F. Hubele 2006. *Engineering Statistics*, 3rd Ed. John Wiley & Sons, New York, NY.

Moubray, J. 1997. *Reliability-Centered Maintenance*, 2nd Ed. Butterworth Heinemann, Oxford, U.K.

Murphy, K.E., C.M. Carter, & R.H. Gass. 2003. "Who's Eating Your Lunch? A Practical Guide to Determining the Weak Points of Any System." Proceedings of the Reliability and Maintainability Symposium, Tampa, FL.

Murphy, K.E., C.M. Carter, & A.W. Malerich. 2007. "Reliability Analysis of Phased-Mission Systems: A Correct Approach." Proceedings of the Reliability and Maintainability Symposium, Orlando, FL.

Nabhan, M.B. 2010. "Effective Implementation of Reliability Centered Maintenance." *AIP Conference Proceedings*, 1239, 88–95.

Nachlas, J.A. 2005. *Reliability Engineering: Probabilistic Models and Maintenance Methods.* Taylor & Francis, Boca Raton, FL.

Nowland, S., & Heap, H. 1978. *Reliability Centered Maintenance*. Dolby Access Press.

O'Connor, P.D.T. 2002. *Practical Reliability Engineering*, 4th Ed. John Wiley & Sons, New York, NY.

Pipe, K. 2008. "Engineering the Gateway for Implementing Prognostics in CBM." Proceedings of the American Helicopter Society Specialists Meeting on Condition Based Maintenance, Huntsville, AL.

Pukite, J., & P. Pukite 1998. *Modeling for Reliability Analysis*. IEEE Press, New York, NY.

Ramakumar, R. 1993. *Engineering Reliability Fundamentals and Applications*. Prentice-Hall, Upper Saddle River, NJ.

Relex Software Corporation. 2003. *Reliability: A Practitioner's Guide*. Intellect.

Sage, A.P. 1992. *Systems Engineering*. John Wiley & Sons, New York, NY.

Sautter, F.C. 2008. "A Systems Approach to Condition Based Maintenance." Proceedings of the American Helicopter Society Specialists Meeting on Condition Based Maintenance, Huntsville, AL.

Shanley, F.R. 1967. *Mechanics of Materials*. McGraw-Hill Book Co., New York, NY.

Shigley, J.E. 1977. *Mechanical Engineering Design*, 3rd Ed. McGraw-Hill Book Co., New York, NY.

Sillivant, D., & Farrington, S. 2012. "Determining the Availability on a System of Systems Network." In *Reliability and Maintainability Symposium 2012*. Reno, NV, pp. 1 –5.

Singh, J., S. Vittal, & T. Zou. 2009. "Modeling Strategies For Reparable Systems Having Multi-Aging Parameters." Proceedings of the Reliability and Maintainability Symposium, Ft Worth, TX.

Smith, A.M. 1993. *Reliability-Centered Maintenance*. McGraw-Hill, New York, NY.

Snook, I., J.M. Marshall, & R.M. Newman. 2003. "Physics of Failure as an Integrated Part of Design for Reliability." Proceedings of the Reliability and Maintainability Symposium, Tampa, FL.

Tebbi, O., F. Guerin, & B. Dumon. 2003. "Statistical Analysis of Accelerated Experiments in Mechanics Using a Mechanical Accelerated Life Model." Proceedings of the Reliability and Maintainability Symposium, Tampa, FL.

van den Bogaard, J.A., J. Shreeram, & A.C. Brombacher. 2003. "A Method for Reliability Optimization through Degradation Analysis and Robust Design." Proceedings of the Reliability and Maintainability Symposium, Tampa, FL.

Vaughan, R.E., & D.O., Tipps. 2008. "A Condition Based Maintenance Approach to Fleet Management 1." Proceedings of the American Helicopter Society Specialists Meeting on Condition Based Maintenance, Huntsville, AL.

Wang, W., & J. Loman. 2003. "A New Approach for Evaluating the Reliability of Highly Reliable Systems." Proceedings of the Reliability and Maintainability Symposium, Tampa, FL.

Warrington, L., & J.A. Jones. 2003. "A Business Model for Reliability." Proceedings of the Reliability and Maintainability Symposium, Tampa, FL.

Weibull.com. 1992–2014. "Reliability Engineering, Reliability Theory and Reliability Data Analysis and Modeling Resources for Reliability Engineers." Available at: http://www.weibull.com/

Wessels, W.R. 2003. "Cost-Optimized Scheduled Maintenance Interval for Reliability-Centered Maintenance." Proceedings of the Reliability and Maintainability Symposium, Tampa, FL.

Wessels, W.R. 2004. "Reliability-Centered Maintenance for Mining Machinery." Proceedings of the 2004 MINExpo, National Mining Association, Las Vegas, NV.

Wessels, W.R. 2007. "Use of the Weibull versus Exponential to Model Part Reliability." Proceedings of the Reliability and Maintainability Symposium, Orlando, FL.

Wessels, W.R. 2008. "Reliability Functions of Flight-Critical Structural Materials from Stress-Strength Analysis." Proceedings of the American Helicopter Society 64th Annual Forum, Montréal, Canada.

Wessels, W.R. 2010. *Practical Reliability Engineering and Analysis for System Design and Life-Cycle Sustainment*. CRC Press, Boca Raton, FL.

Wessels, W.R. 2011. "Mechanical Engineering Design-for-Reliability." Proceedings of the 2011 International Mechanical Engineering Congress & Exposition, Denver, CO.

Wessels, W.R. 2012. "Reliability Engineering Approach to Achieve RCM for Mechanical Systems." Proceedings of the 2012 Reliability and Maintainability Symposium, Reno, NV.

Wessels, W.R., W. Roark, & S. Hardy. 2004. "Application of Modeling and Simulation to Predict Impact on System Availability due to Logistical Decisions for Sparing and Resource Allocations." Proceedings of the Huntsville Simulation Conference.

Wessels, W.R., & F.C. Sautter. 2008. "Reliability Functions of Flight-Critical Structural Materials from Stress-Strength Analysis." Proceedings of the International American Helicopter Society Forum, Montreal, Quebec, Canada.

Wessels, W.R., & F.C. Sautter. 2009. "Reliability Analysis Required to Determine CBM Condition Indicators." Proceedings of the Reliability and Maintainability Symposium, Ft Worth, TX.

Wolstenholme, L.C. 1999. *Reliability Modeling A Statistical Approach*. Chapman & Hall/CRC, London, U.K.

Xing, L., P. Boddu, & Y. Sun. 2009. "System Reliability Analysis Considering Fatal and Non-Fatal Shocks in a Fault Tolerant System." Proceedings of the Reliability and Maintainability Symposium, Ft Worth, TX.

Zhang, Y., R. Rogers, & T. Skrzyszewski. 2003. "Reliability Prediction of Hydraulic Gasket Sealing." Proceedings of the Reliability and Maintainability Symposium, Tampa, FL.

Index

A

Achieved availability, 9
 system sustainment and, 120
Active parallel design configuration,
 54–55
Affordability, reliability, 19
Allocation
 part reliability, 49–51
 reliability, 251–252
American Society of Mechanical
 Engineers, 6
Amounts *versus* equivalent value, 126
Assembly and system simulations,
 268–270, 271–273
Assembly reliability functions, 53–63
Availability, 5, 8–10
 part design and, 84–86
 system sustainment and, 119–120

B

Baseline reliability analysis, 90–91

C

Capital, operating, 124
Capital costs, 169–173
Capital recovery, 334–338
 loan, 135
Cash flow timelines, 130
 capital costs in, 169–173
 hourly rate estimation, 174–175, *176*
 lost opportunity costs, 176–177
 m=1, 134–139
 m=4, 139–142
 net cash flow, 174
 operating and maintenance costs in,
 173–174
 for three projects: m=1, 142–144
Catastrophic consequences, 38

Classifications of sources and uses of
 estimated cash, 127–129
Complete data
 in exponential reliability math
 model, 96–98
 Weibull reliability math models,
 101–103, *104*
Compounding periods, 126
 m=1, 134–139
 m=4, 139–142
Condition-based maintenance (CBM),
 29–30, 233, 235–236
Confidence limits, 148–150
 deterministic reliability analysis,
 183–184
Consequences, 39
 analysis, 38–39
Corrective and preventive maintenance,
 154–155
Cost estimation, 126–127
 cash flow timelines and, 130
 classifications of sources and uses of
 estimated cash from, 127–129
 future amount, 325–326
 present amount, 325
 uniform recurring amounts, 328–330
Cost objective function, 157–163
Critical items list (CIL), 21–22
 part, 35–41
 qualitative failure, repair, and
 logistic analysis, 22–25
 quantitative failure, repair, and
 logistic analysis, 25–31
Criticality analysis, 19–22
 modified Moubray, 37–39
 proposed procedure for, 41, *42*
 rank, 40–41
 reliability database and, 246–247
Cumulative failure probability
 distribution, 53